Ancient Greek and Roman Science: A Very Short Introduction

VERY SHORT INTRODUCTIONS are for anyone wanting a stimulating and accessible way into a new subject. They are written by experts, and have been translated into more than 45 different languages.

The series began in 1995, and now covers a wide variety of topics in every discipline. The VSI library currently contains over 700 volumes—a Very Short Introduction to everything from Psychology and Philosophy of Science to American History and Relativity—and continues to grow in every subject area.

Very Short Introductions available now:

Liba Taub

ANCIENT GREEK AND ROMAN SCIENCE

A Very Short Introduction

Great Clarendon Street, Oxford, OX2 6DP,
United Kingdom

Oxford University Press is a department of the University of Oxford.
It furthers the University's objective of excellence in research, scholarship,
and education by publishing worldwide. Oxford is a registered trade mark of
Oxford University Press in the UK and in certain other countries

Published in the United States of America by Oxford University Press
198 Madison Avenue, New York, NY 10016, United States of America

British Library Cataloguing in Publication Data

Data available

Library of Congress Control Number: 2022947357

ISBN 978-0-19-873699-8

Printed and bound by
CPI Group (UK) Ltd, Croydon, CR0 4YY

For Niall

Contents

Ancient Greek and Roman Science

Acknowledgements

It is a pleasure to thank a number of friends, colleagues, and former students for their help in preparing this volume for the *Very Short Introduction* series. Given that it is so short, it took a surprisingly long time to complete, and I am very grateful to the following, as well as other friends and colleagues who have supported and humoured me over the years: Peter Adamson, Terri Apter, Jude Browne, Paul Cartledge, Aude Doody, Llinos Edwards, Seb Falk, Robin Lane Fox, Arthur Harris, Nick Jardine, Alexander Jones, Sachiko Kusukawa, Joe Martin, Mary Jo Nye, Marilyn Ogilvie, Charlie Pemberton, Emma Perkins, Rachel Rowe, Tom Ryckman (for Schilpp quotation), Maartje Scheltens, Ineke Sluiter, Laurence Totelin, and Frances Willmoth. At Oxford University Press, Latha Menon, Jenny Nugee, Imogene Haslam, and Meredith Taylor were exceedingly patient and helpful, as was Saraswathi Ethiraju at Straive. I also thank the anonymous readers for their comments. Newnham College was generous in support of my work, as was the Department of History and Philosophy of Science. I am especially grateful to Konrad Bieber, for giving me his copy of Erwin Schrödinger's *Nature and the Greeks*, very many years ago. And above all, I thank Niall Caldwell, for so much.

List of illustrations

Chapter 1
Understanding the world

Science is at the centre of modern society. Whoever we are, we probably look to science for answers, at least for some things. If something is 'scientific', we may think that it is more reliable or better than something that is not. This, in part, accounts for the widespread interest in the history of science. The origin and early history of scientific ideas and practices hold a special fascination.

Ancient Greeks are often regarded as having invented science. This is because some ancient Greeks explained natural phenomena without attributing them to supernatural causes. They have also been credited by some with inventing the idea of nature. This volume in the *Very Short Introduction* series concentrates on attempts to understand the natural world, rather than strategies to control it. Hence it is primarily concerned with science rather than technology, and gives a brief, roughly chronological account of ancient scientific ideas and practices covering a broad sweep of time, from the 8th century BCE to the 6th century CE, and focusing primarily on Greece and Rome.

The ancient natural philosophers—lovers of wisdom concerning nature—sought to explain the order and composition of the world, and how we come to know it. They were particularly interested in what exists (ontology) and how what exists is ordered (cosmology). They were also concerned with how we come to know (epistemology) and how best to live (ethics). Philosophers today—and many others—are still fascinated by questions of what exists and how we know.

The English word for nature comes from the Latin *natura*; *physis* is the Greek word translated as 'nature', and the word 'physics' derives from this. Both words, *physis* and *natura*, have been studied in detail, and histories of their usage show that each had a range of meanings and a variety of things to which they referred, and that these were not static. There was no single shared conception of 'nature' among ancient Greeks, and the investigation of nature was part of a larger culture of enquiry. For many early Greek philosophers, the study of nature focused on explaining what we see (phenomena). For some, 'nature' included the human body and disease. And, not only animals, humans, and plants but the cosmos itself was understood to be alive. The Greek noun *kosmos* means 'order' or 'adornment', and it is the origin of the English word 'cosmetic'. Probably by the 5th century BCE, the word *kosmos* had taken on the meaning of 'world order' and even 'world', conveying the idea of an apparent order displayed through the motions of the Sun, Moon, and stars.

Modern accounts of ancient science sometimes emphasize certain perceived features, which are implicitly, if not explicitly, approved of or disparaged: the rejection of gods and myth (good); the importance of rational explanation (also good); an emphasis on mathematics (excellent); a lack of observation and experimentation (bad); the practice of pseudo-science, including astrology (very bad). In other words, ancient science is often viewed through a modern lens, and graded according to an anachronistic vision of what is now considered to be scientific theory and methodology. There is also often an assumption that science was only a Greek endeavour, that the Romans did not 'do' science.

Yet, the evidence from the period itself challenges many of these perceptions and judgements. For example, there are records of extensive and systematic observations of different features of the natural world (living beings as well as astronomical phenomena) and of various scientific instruments being devised and used. Some Romans (writing in Latin) did engage in scientific work, offering explanations of natural phenomena. Looking back to their intellectual forebears, Roman thinkers did not always accept the Greeks' views, and sometimes criticized their approaches. Many ancient Greek and Roman scientific works invoke or refer to something divine (eternal), including the cosmos itself; some accounts—even those which are deemed to be 'mathematical'—have quasi-religious connotations. The organization and relative value of different kinds of knowledge (including mathematics) were debated. Engagement with prior authorities—such as the earliest Greek epic poets, Homer and Hesiod—was complicated: tradition was sometimes celebrated, but so was innovation and discovery.

Ancient philosophers were not the only ones who loved wisdom about nature, nor were mathematicians the only ones fascinated by numbers, counting, and calculating. All sorts of ancient texts—including poetry, drama, and history—contain scientific ideas, arithmetic, and geometry. Questions about nature, the human body, and even mathematical problems were the stuff of dinner-party conversations for educated men. What are nowadays marked as specifically scientific concerns permeated across ancient society and culture. Then, as now, scientific work was not done in a vacuum, and other cultural values came into play—including those associated with art, religion, and politics. And there was a continual questioning of how much can really be known—not just about nature, but about anything.

Today, there are many questions about how science is done. Who is doing science, and why? Who has a stake in the results? Who pays for scientific work, and who benefits from it? We can ask the same questions about science in the ancient world. Indeed, in antiquity there was persistent interest in addressing questions about how best to 'do' natural philosophy and mathematics, including querying what counts as the 'right' approach. This sort of interrogation highlights the ambition of many of those whose work survives. We also encounter questions—as well as statements—about motives for pursuing science and mathematics: what are science and mathematics good for? What are the benefits or pay-offs? Furthermore, a number of important medical authors, including the 2nd-century CE physician Galen of Pergamum, advocated that the best doctors are also philosophers. The disciplinary boundaries between modern intellectual specialisms did not apply in antiquity.

As today, certain individuals in antiquity gained great reputations for their scientific work. When we look at the names associated with scientific ideas and achievement in antiquity, we see that some people were pursuing science for a gain that was very specific, but not financial. In the 4th century BCE, Plato advocated the study of mathematics as part of the training to be a philosopher-king, to guide the *polis* (the city-state) and its citizens. This knowledge of mathematics was not intended to be 'knowledge for its own sake', but knowledge for the purpose of ruling. For some, a primary motivation to pursue science was to alleviate fear of the unknown, by offering a rational account of potentially scary phenomena without having recourse to gods who acted irrationally. It is not clear what sort of public reward or recognition was possible during their lifetimes, even for the very distinguished. Some scientists and mathematicians may have had patrons, but it is difficult to say to what extent such patronage relationships normally included financial benefit (and in what form).

Relationships between teachers, students, and followers were important in ancient Greece and Rome. Intellectual as well as cultural connections mattered, as did heritage. We have clear references to individuals active in the 6th century BCE and their associates, including Thales and those who 'heard' him (presumably his students), as well as Pythagoras and his followers, who unusually included women, some of whom we know by name. In later periods, allegiance to philosophical schools was emphasized. 'Schools' provided the setting and identity for a great deal of work in natural philosophy. These schools of thought or sects (*hairesis* = sect) were typically associated with a particular founding figure and approach; the word *hairesis* also has the meaning 'choice', suggesting that members had chosen affiliation with a particular school. Some individuals—including founders of schools—were famous in antiquity for their scientific work, becoming 'heroes' or demi-gods.

The work of celebrated individuals was sometimes displayed on expensive, inscribed stone monuments serving almost as a precursor to modern billboards, marking achievement and sharing information publicly. Scientific work was communicated through a range of media, but our historical evidence for the study of ancient science is, for the most part, to be found in written texts. There is some surviving archaeological evidence, notably hundreds of sundials, a famous device known as the Antikythera Mechanism, a globe or two, stone calendars correlating weather with astronomical events, and wind-roses indicating wind direction. We also have extensive visual material conveying relevant information and ideas: painted vases, frescos, mosaics, and coinage. Furthermore, we have ancient accounts of various scientific instruments that no longer exist. While much of what follows here is based on written evidence, material and visual sources are of distinct value.

Aristotle wrote that philosophy—which would include what we call science—was the most pleasant activity. However, we don't actually find many clear statements from Greek or Roman authors that they pursued scientific activities because they were inherently interesting or enjoyable. A notable exception is the Roman statesman, author and philosopher Lucius Annaeus Seneca ('the Younger', active in the 1st century CE), who while in exile wrote to his mother explaining that:

> …these days are my best, because my mind is relieved from all pressure of business and is at leisure to attend to its own affairs, and at one time amuses itself with lighter studies, at another eagerly presses its inquiries into its own nature and that of the universe: first it considers the countries of the world and their position: then the character of the sea which flows between them, and the alternate ebbings and flowings of its tides; next it investigates all the terrors which hang between heaven and earth, the region which is torn asunder by thunderings, lightnings, gusts of wind, vapour, showers of snow and hail. Finally, having traversed every one of the realms below, it soars to the highest heaven, enjoys the noblest of all spectacles, that of things divine, and, remembering itself to be eternal, reviews all that has been and all that will be for ever and ever.

Two other Roman authors roughly contemporary with Seneca, Pliny the Elder and Columella, suggest that much of the motivation for engaging in scientific activity was to have some sort of control over nature (as agriculture and medicine aim to do), and also to harness the power of nature for the present and for posterity. The distinction between science and technology is not always clear-cut: in medicine, some physicians were interested in understanding health and disease, even while their primary goal was to control both.

A fascination with other peoples and cultures is found in the writings of various Greek and Roman authors. Seneca reports the ideas of the Chaldeans (in Mesopotamia; often associated by Roman writers with astronomy), and Pliny takes pride in listing his non-Roman—including Carthaginian—sources. Many natural philosophers travelled to other places for the purpose of study. Thales of Miletus, for example, was reputed to have gone from there to Egypt, where he spent time with priests. Diogenes Laertius, who probably wrote his *Lives of Eminent Philosophers* in the first half of the 3rd century CE, refers specifically to philosophers in other lands, including the Gymnosophists (the so-called 'naked wise men') of India. Both Greeks and Romans grappled with data and explanations of the natural world emanating from other cultures and linguistic groups. Some Roman authors sought to make Greek ideas available in Latin, even while others were determined to demarcate between the two cultures.

The focus in this volume is on ancient Greeks and Romans (Figure 1). Their ideas and practices are worth studying for their own sake. They also provided the foundation upon which Western scientific traditions developed. Other ancient cultures—including Babylonian, Egyptian, Indian, and Chinese traditions—also sought to explain the world and increase knowledge. Some of these endeavours can be seen as scientific, philosophical, and/or mathematical, and a number of them influenced ancient Greeks and Romans. Bearing in mind the ancients' own interest in and concerns with other peoples, cross-cultural influence is part of the story of ancient science.

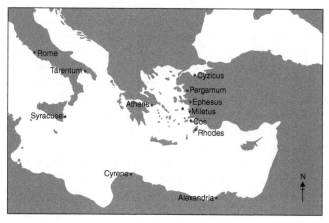

1. **Notable locations.**

Ancient Greece and ancient Rome

The geographical, cultural, and political boundaries of 'ancient Greece' and 'ancient Rome' were not static. When we refer to 'ancient Greece', we are pointing to a civilization dating from about the 12th century BCE until the end of 'antiquity', c.600 CE, that at some points in time included not only what is geographically modern Greece, but also the Black Sea rim, parts of western Asia, the Italian peninsula, Sicily, eastern Spain, the south of France, and north Africa. Politically, several hundred relatively independent city-states (*poleis*, plural of *polis*) identified as Hellene (i.e. Greek), sharing various cultural features including language (even if the dialect varied) and religion (once again, with variations). The mountainous geography, interlaced with rivers, contributed to a lack of cohesion. A number of *poleis* set up colonies, some of which were very far away. Colonies were independent yet linked to their 'mother' city-state in various ways, for example, by a shared calendar.

'Ancient Rome' refers to the civilization that grew from the founding of the city of Rome in the 8th century BCE, according to legend, by members of the Latin tribe. The Latin language was originally spoken in the area around Rome, known as Latium. Due to the power and influence of Rome, Latin became the dominant language in the Italian peninsula, and eventually throughout the western Roman Empire, which geographically included parts of north Africa, modern Europe, and Britain as well as the Mediterranean region and portions of western Asia. For over a thousand years, Greeks and then Romans (first through the Republic and then the Empire) dominated the Mediterranean area. The Hellenistic period refers to the time from the death of Alexander the Great in 323 BCE to the emergence of the Roman Empire (following the Battle of Actium in 31 BCE and the conquest of Ptolemaic Egypt in 30 BCE).

Some Romans adopted aspects of Greek culture. For this reason, the modern term Graeco-Roman is often used to highlight shared aspects of culture among ancient Greeks and Romans. Here, because so many of our historical sources are written texts, the original language of authorship is important, to some extent signalling an author's self-identification as culturally Greek or Roman. But some of the 'actors' in our story were bilingual, even translating Greek material into Latin for a Roman audience, simultaneously underscoring both shared interests and important differences.

Chapter 2
Expert poets

The oldest known Greek poets, Homer and Hesiod—credited with authorship of the *Iliad* and the *Odyssey*, and the *Theogony* and *Works and Days* respectively—were often regarded in antiquity as experts on all sorts of topics, including astronomy and weather, the gods, human behaviour, and morality. One ancient account of their expertise, *The Contest of Homer and Hesiod*, describes a riddling game featuring numbers and counting; it is the sort of thing that some of us might remember from our younger days as a 'story problem'. The character called 'Hesiod' posed the following question: how many Achaeans (Greeks) went to Troy together with the sons of Atreus? The other character, called 'Homer', answered with a counting problem: there were 50 hearths, and in each one there were 50 spits, and on each spit 50 pieces of meat, and three (times) 300 Achaeans were around each piece of meat. The narrator—or possibly a later (Byzantine) annotator—attempted the calculation: there may have been 112,500,000 men.

The *Contest* itself is about who is the better poet, but not only that: the two poets also show off their mathematical prowess. Indeed, this 'story problem' is set against the background of the Trojan War, sending a strong signal that mathematical brainteasers were embedded within broader Greek culture, and that at least some people thought it was entirely possible to be

'Homer' and 'Hesiod'

When we use the names 'Homer' and 'Hesiod', we are using a sort of shorthand to refer to 'authors' about whom we know very little. Since antiquity, the dating of the poems ascribed to Homer and Hesiod has been a topic of discussion. Indeed, the poems were the focus of commentaries by various ancient authors. Many modern scholars judge the Homeric poems (credited to Homer) to be from the 8th century BCE, and the Hesiodic poems (attributed to Hesiod) to be somewhat later, but some date Hesiod first. Furthermore, there have been suggestions that 'Homer' may have been more than one person. In the 1920s and 1930s, Milman Parry and his student Albert Lord argued that the formulaic nature of the Homeric poems would have lent itself to oral composition and transmission; this idea was based on evidence of oral composition in modern Serbian epic poetry. At the end of the 19th century, Samuel Butler argued that 'Homer' was a woman. Few doubt that Hesiod was a real—and single—person.

both a great poet and mathematically able. Such dual expertise in literature and mathematics or science has not always been assumed possible. In the 1950s, the British scientist and novelist C. P. Snow famously criticized the British educational system for acting as if there are 'two cultures', one of literary intellectuals, the other of scientists. The term 'Renaissance man' has been used to refer to someone who has expertise in several areas, including the arts as well as science. The poets we call 'Homer' and 'Hesiod' would probably have qualified as 'Renaissance men', about 2,000 years before the Renaissance. There was no sign of 'two cultures' in ancient Greece and Rome.

Standing at the fountainheads of tradition, the epic poets helped shape intellectual agendas. Plato and Aristotle both suggested, perhaps jokingly, that Hesiod and Homer were the fathers of philosophy. While they were revered for their ideas, both were credited with great practical knowledge too. Hesiod was named by later Greeks and Romans as the father of agriculture. Aristotle treats Homer—who lived hundreds of years earlier—as a well-respected authority on the names of places, even those as far away as Egypt.

The epic poems did not focus on the explanation of nature. The Homeric *Iliad* is concerned with the Trojan War and its consequences: war, peace, gods, mortals, and human and divine heroes feature throughout. The *Odyssey* recounts the travels of Odysseus, one of the heroes of the *Iliad*. He visits strange and mythical places as well as real ones; gods and humans have equally active roles. Hesiod's *Theogony* describes the world and its coming to be; his *Works and Days* discusses some technical matters, including agriculture. All of these poems convey information about the natural world—including the seasons and constellations—relevant to 'science'; this information was taken extremely seriously, over a very long period of time.

Star-gazing

The sense of an annual cycle of recurrent seasons is present in the Homeric poems ('with the year's full circling the seasons returned'), but there is only a rough division of the year into seasons. The poet recognized that certain stars are more visible at different times, serving as the basis for reckoning the time of year. Seasons are characterized by particular sorts of weather: winter is a time of unceasing rain and storms; spring is windy and also the time of the nightingale's song; and late summer/early autumn is the season of harvest and falling leaves, violent downpours, and flooding. Stars were sometimes seen as portents of catastrophe. Late summer/early autumn was referred to as the time of the star known as Dog-of-Orion (Sirius), and associated with fever.

Vividly drawn passages in the Homeric poems give an indication of contemporary knowledge and ideas about astronomical phenomena. At one point in the *Odyssey*, a wakeful Odysseus sails from the goddess Calypso's island, carefully keeping his eye on constellations in the night sky. This passage is often pointed to as an illustration of Homer's practical knowledge of celestial navigation. Following Calypso's instructions, Odysseus kept the Bear (which we call Ursa Major) on his left and used the stars to help him navigate:

> Cheered by this breeze glorious Odysseus spread his sail,
> and sitting next to the steering oar kept the raft skilfully
> on course; nor did sleep ever fall on his eyelids, but he
> held his gaze on the Pleiades and late-setting Boötes,
> and the Bear that men also call the Wain, which turns
> always in the same place and keeps close watch on Orion,
> and alone has no share in the baths of Ocean; for indeed
> Calypso, bright among goddesses, had instructed him
> to keep this star on his left as he sailed across the sea.

This may be the only place in the Homeric poems where the stars are used in this way. The Bear is singled out as neither rising nor setting, but always visible. Here we have a brief glimpse of knowledge about astronomical bodies prior to the 6th century BCE, when we get a sense of work being done that might be called 'astronomy'. As the audience for these poems was unlikely to have been especially interested in studying the night sky, the phenomena mentioned—the named constellations—were probably well known. The stars and the Sun served as direction-markers.

At the same time we are reminded of the presence of the gods, for it is not only the stars that Odysseus relies on to navigate, but also the goddess Calypso. The Sun is also a god (Helios) and the Moon a goddess (Selene). In the Homeric and Hesiodic poems we find that many things are in the control of, or linked to, individual gods. In the *Odyssey*, Helios owns seven herds of 50 cattle and seven herds of 50 sheep. The herds of Helios are special, completely unlike the herds tended by humans, for they neither increase nor diminish in number. The Sun's cattle described by Homer reappear (and are repurposed, for example, in arithmetical problems) in later Greek scientific and mathematical writings. Aristotle and others read the reference to Helios' herds allegorically, and understood this description of the property of the Sun as referring to day and night, suggesting that the 350 cattle and 350 sheep represented the days and nights of the lunar year (which is about 354 days long).

Stormy weather

Specific winds were also thought by some to be gods, or to be associated with gods. In the Hesiodic poems, Boreas (the north wind), Notos (the south wind), and Zephyros (the west wind) are described as offspring of the god Astraios and goddess Eos, while a fourth wind, Euros (the east wind), also features in the Homeric poems, along with other unnamed winds. Writing in the 5th century BCE, the historian Herodotus explained that Boreas was a minor god who intervened against the Athenians' naval enemies and was, in consequence, worshipped. In antiquity, painted vases depicted Boreas with wings, indicating his speed (Figure 2). Aristotle twice mentions painters' portrayals of wind, specifically Boreas.

Not only are the winds personified, there are also stories about them. The *Odyssey* emphasizes their importance for sailing. The god Zeus makes Aiolus, a mortal, responsible for all of them, secured in a leather pouch; he released only the west wind to enable ships to be blown safely on their journeys. When some of Odysseus' men, hoping that Aiolus' bag held treasures, untied the string closure, the danger of wind was exposed: they let loose all the winds, causing a terrible storm that swept them away.

2. **Boreas, the north wind, featured on a 5th-century BCE wine jug made in Attica.**

The use of mythology to describe and explain what we today regard as 'natural' phenomena was important to the fabric of Graeco-Roman culture. Many of these phenomena, including storms, lightning, and thunder, are potentially dangerous and frightening. Traditional mythology offered explanations of meteorological events as acts or epiphanies of gods. In the *Iliad* and *Odyssey*, Zeus, sometimes referred to as 'cloud-gatherer', is often responsible for weather events. He produces thunder and hurls bolts of lightning, causing rain and storms. He also places rainbows in the clouds as a portent for humans. Other divinities can also produce meteorological phenomena: together, Hera and Athene cause thunder. Euros and Notos raise the waves, springing from the clouds of Zeus. Winds can also be controlled by various gods and goddesses, including Athene and Calypso. Poseidon, often described as 'earth-shaker', has control not only over seismological but also meteorological phenomena, pulling together clouds and producing wind storms.

While Homer provided vivid descriptions, Hesiod stands at the beginning of a long tradition of predicting the weather, often using astronomical phenomena. He was reputed in antiquity to have written a now-lost work entitled *Astronomy*, and in his poem the *Works and Days* we learn that knowledge of the motions of the astronomical bodies is useful for weather prognostication. The final section reads like an early farmers' almanac, the kind still produced in the United States annually and which still claims to be based on astronomical knowledge. Today such almanacs are available online as well as in print; Hesiod's was presented as a poem, injected with practical tips for his (feckless) brother Perses. He offers advice for steering clear of various calamities. Work, especially agriculture, helps us to avoid many bad things. Hesiod provides practical maxims for everyday life ('do not postpone things until tomorrow'), as well as a list of the best days for work.

Familiarity with astronomical events enables us to know which times of year are appropriate for various tasks:

> When Orion and Sirius come into mid-heaven, Perses, and rose-fingered Dawn meets Arcturus, then set about cutting off all the grape-clusters for home .... But when the Pleiades and Hyades and mighty Orion are setting, then be thinking of ploughing.

The poet does not simply give general advice for using the stars to tell the seasons but also presents a detailed calendar that specifies when particular activities should be undertaken.

Agricultural tasks are linked to what can be observed in the sky, the risings and settings of bright stars, and the apparent motion of the Sun. Hesiod advises:

> when the Pleiades, running before Orion's grim strength, are plunging into the misty sea, then the blasts of every kind of wind rage; at this time do not keep ships on the wine-faced seas, but work the earth.

Hesiod also looks to birds, and suggests that the annual cry of the crane signals the beginning of planting and winter rains. It is unlikely that he learned all of this on his own. While he probably picked up a good deal on the family farm, the level of detail suggests that—probably over a lengthy period of time and through shared effort—Boeotian farmers had devised a calendar to guide their agricultural work throughout the year. Here the poet shares the fruits of this communal labour, presumably passed from generation to generation.

Out from Chaos: the cosmos

Hesiod was not interested only in practical matters. His thoughts turned to the cosmos itself in his *Theogony*, where he offers a genealogy of the gods. They include personified cosmic beings: Chaos (the Chasm) was born first, followed by Earth, the Sky, the inner Sea, and outer Ocean. Hesiod's poem is a cosmogony, an account of the origin of the cosmos. In it, the births of the gods give rise to the main physical features of the world. And, while Chaos featured prominently here, the idea of cosmic order became key to ancient Greek and Roman understandings of the world.

Some modern scholars discount mythological accounts as being of no interest from a 'scientific' standpoint. Such a view has a distinguished history: Aristotle, in his *Poetics*, may have been the first to contrast poets recounting myths about gods with philosophers investigating the natural world. But in spite of the clear distinction that Aristotle made between these two approaches, other ancient philosophers suggested that the relationship between mythology and philosophy was more complicated. Plato chose to present some cosmological and geographical information in the form of myth. Some ancient readers adopted rationalizing interpretations of the epic poems. Writing in the 6th century BCE, Theagenes of Rhegium suggested that Homer spoke allegorically rather than literally in the *Iliad* when he identified gods with the natural elements, and hot and cold, dry and moist. The distinction between the 'rational/scientific' and the 'mythic' was not always clear-cut.

The development of Greek science is sometimes seen as a rejection of the traditional myths embedded in Homeric and Hesiodic poetry. But myth was not entirely discarded in subsequent scientific accounts in antiquity; nor was tradition, including the poetic tradition. Poetry was a principal genre of communication on a range of subjects, indicating the power and authority conveyed by epic metre, because of the association with Hesiod and Homer. It was an especially powerful format for ancient Greek and Roman philosophical and scientific ideas. Many later ancient Greek and Latin poets—including Aratus in the 3rd century BCE, Lucretius and Virgil in the 1st century BCE, and Manilius in the 1st century CE—adopted epic metre for poems that offered accounts of the physical world. In doing so, these later poets associated themselves with the authority and cachet of the epic tradition, while sharing scientific information and explanations.

Respect for earlier thinkers can be seen in the work of many later authors writing on scientific and mathematical topics. While the epic poets may not have had the sharing of scientific and technical information as their primary aim, portions of their work are quoted or alluded to by many later authors writing on these subjects, to illustrate and support their own arguments, and to show off that they know their poets. Pliny the Elder considered Hesiod to be as valid an authority on astronomical matters as specialist astronomers. Many ancient authors quoted Homer as a source of knowledge about the natural world. Diogenes Laertius reports that Metrodorus of Lampsacus (active in the 5th century BCE) was the first to occupy himself with Homer's physical ideas. The large number of references to the Homeric and Hesiodic poems in the works of later ancient Greek and Roman authors—even prose works on specialized, technical subjects—demonstrate that the epic poets had great intellectual impact, even on topics for which we would not normally expect poets to be experts.

The American classicist David Konstan has suggested that, in concentrating on the mythic genealogy of the gods in the *Theogony*, many of whom have important interactions with the physical world, Hesiod left room, so to speak, for the materialist explanations of the world offered in the 6th and 5th centuries BCE. Hesiod's Chaos is at the start of the conception of 'cosmos' and its order. Providing an account of this world order—via cosmology, the study of the structure of the world as a whole—was a principal preoccupation of many philosophers interested in nature. There was a presumption that the cosmos was ordered and beautiful, and also the view that this beauty did not refer only to a physical attribute but also to ethical virtue. And, there were still other important questions to consider: what is responsible for the order and structure of the world? What is it made of? How can we explain various things we observe? What can we know, and how?

Chapter 3
Inventing nature

Aristotle referred to many of his predecessors as *physiologoi*, those who theorized about nature and attempted to answer fundamental questions about the world and how we understand it: what is the world made of? Is there some primary matter or stuff? What really exists? The *physiologoi* (of whom Aristotle named Thales of Miletus as the first) are regarded primarily as philosophers, probably because most of what we know about them has come to us via ancient accounts of the history of philosophy. While the *physiologoi* were famous for their ideas, and today we tend to emphasize their status as philosophers, in antiquity their practical pursuits were also well known and celebrated. We have biographies of a number of them, providing details of their lives, work, ideas, interests, and motivations, but only fragments of their actual writings, when we have them at all. We also have reports of their hands-on activities and inventions—especially those related to astronomy and meteorology—from writers interested in such topics, including the 1st-century BCE Roman architectural author Vitruvius, who credits by name those who designed specific kinds of sundials. These achievements are as important as their ideas, as we will see.

The earliest Greek philosophers are often called 'Presocratic', but the term is not precise and some of the 'Presocratics' were actually contemporaries of Socrates, who died in 399 BCE. Many ancient

Fragments and testimony

It is hard to say much about the ideas and work of the early Greek philosophers with confidence because their writings (for the most part) do not survive. However, we do have references to their ideas in writings by other (sometimes much later) ancient authors. The reporting and discussion of earlier ideas and activities by later thinkers demonstrate their importance: if they hadn't been thought worth saving and repeating, people wouldn't have bothered. The philosopher and commentator Simplicius, who lived in the 6th century CE and is a source about the ideas of many earlier thinkers, explained that he had copied out a portion of Parmenides' poem, composed in the 5th century BCE, because copies of it were so rare.

Direct quotations are referred to as 'fragments', while a report or paraphrase is called 'testimony'. Those fragments and testimonies that survive give a sense of the ideas attributed to their predecessors by later ancient authors, who wrote them down, preserved, and circulated them. Since much of our knowledge is second-hand, at best, the source of our information about individual early philosophers must be considered, as well as the amount of time—sometimes centuries—that had passed between the original thinker and the report. Knowing the source may alert us to possible prejudice (favourable or not) on the part of the individual sharing the information. For example, Aristotle often surveys the opinions of others before presenting his own—in his view, superior—ideas.

Greek philosophers lived in regions that today are not 'Greek'. For example, Thales, Anaximander, and Anaximenes all lived in 6th-century BCE Miletus (in present-day Turkey), the Pythagoreans in what is now Italy. Some travelled to various parts of the world, enabling the transmission of ideas, information, and material goods. Herodotus claimed that Greeks had learned about the gnomon (the shadow-caster in sundials, originally simply a rod perpendicular to the Earth) from Babylonians.

The *physiologoi* took part in different activities as part of the study of nature: writing texts, teaching, designing scientific instruments (such as sundials), and making observations and predictions of different phenomena (often astronomical, meteorological, or medical). Some created illustrations of their ideas, including maps and possibly diagrams. The *physiologoi* not only offered new ideas, they invented new modes of communicating about the world.

The *physiologoi* were renowned for their ideas about nature—sharing them, talking about them, and writing them down. We have a sense of ongoing conversations about the topics that they found fascinating, and can imagine that these philosophers had the ideas of others in mind, suggesting a sort of dialogue or even a debate. We know the names of many individuals described in antiquity as philosophers, but for some of these names we have very little other information. Not all of the *physiologoi* appear to have produced written work although most did, and many were said to have written works with the title *On Nature*. However, in almost all cases we must rely on reports from later ancient authors for our knowledge of their ideas and writings, and the titles themselves may have been given by subsequent readers and editors.

In many cases, our only source of biographical detail and the titles of the works of ancient 'scientists' is Diogenes Laertius and his *Lives of Eminent Philosophers*, probably produced in the first half of the 3rd century CE, long after his subjects lived and worked. He is often disparaged as a 'late' source, but he is our sole basis of information for many of the philosophers about whom he wrote. His *Lives* incorporates two approaches to biography: he presents an intellectual history, and is also motivated by an ethical desire to provide exemplars of how to live a good life.

Thales of Miletus

In the 6th century BCE, Miletus was a prosperous and cosmopolitan maritime centre, with numerous colonies throughout the eastern Aegean and Black Sea areas, and many international trading contacts. It was home to a number of *physiologoi* who had a range of interests and practical skills. Thales' astronomical interests feature prominently in accounts of his life. In Plato's *Theaetetus*, Socrates relates that once while out walking at night, Thales was gazing upward, contemplating the stars. Not looking where he was going, he fell into a hole. A Thracian servant-girl saw him and laughed, because he was so keen to know about the things in the sky that he could not see what was at his feet. (This may be an early precursor to absent-minded professor jokes.) Thales was reportedly interested in navigating by the stars, and some claimed that he produced a 'Nautical Star-guide' in verse. But even in antiquity others raised doubts about his authorship. Nevertheless, he was renowned for having successfully predicted an eclipse of the Sun.

Herodotus provides the earliest extant account of that prediction: after five years of warring, the Medes and the Lydians were still battling when day suddenly turned to night. This change had been predicted by Thales, who specified the year in which it would occur. The combatants interpreted the eclipse as a warning of divine disapproval and stopped fighting. Modern historians and astronomers are not sure how Thales made this prediction, if indeed he did. He may have had access to Babylonian eclipse data, based on observation records. The ability to predict may indicate some understanding of a regularity underlying eclipses, or it may not; it is difficult—if not impossible—to know. Herodotus lived about 100 years after Thales, and his report of Thales' prediction may have been widely known. Thales was not thought of simply as an 'armchair' philosopher, but as someone with valuable practical skills. Aristotle also attests to these, noting that Thales once cornered the market on olive presses, using his astronomical knowledge to predict a bonanza harvest and make a good profit.

Aristotle also relates that Thales considered all things to be full of gods. This may be surprising, as the absence of the traditional gods in explanations is often taken as a hallmark of ancient Greek natural philosophy. According to Aristotle, Thales was the first philosopher to say that matter is the original constituent of every thing, identifying water as the primary principle of all things. Aristotle reports that Thales also thought that inanimate objects, such as magnets and amber, had souls which allowed them to move, and that the Earth itself floats on water, like a log. Seneca wrote, disapprovingly, that Thales thought that the world is held up by water, riding like a ship; when it quakes, the rocking is due to the water's motion. It is impossible to know whether Thales really held these views, but there is no reason to doubt it.

Thales' reputation for wisdom was rewarded with prizes. Diogenes Laertius recounts what he described as a well-known story about a tripod, intended to be presented to the wisest person. Actually, he shares several versions, all featuring Thales as the most highly esteemed recipient. Apparently, Thales was the first to receive the title 'Sage' (*sophos*), very special recognition indeed.

Anaximander of Miletus

Like Thales, Anaximander of Miletus (possibly a kinsman and his student) was known for practical wisdom, as well as novel ideas about nature. Anaximander was said to have been the first to draw a map of the Earth and sea, and we can imagine that the maritime character of Miletus shaped his ideas. According to a 3rd-century CE commentator on Aristotle's *Meteorology*, Alexander of Aphrodisias, Theophrastus (working in the 4th century BCE) reported that Anaximander thought that the sea was drying up. Over time, the silting of the Meander River has closed the gulf at Miletus, extending the distance to the coast, and Anaximander may have witnessed part of this process. Today, Miletus is 10 kilometres away from the shore.

Anaximander was credited by some with the invention of the gnomon (despite Herodotus' claims of its Babylonian origins). Diogenes Laertius reports that according to Favorinus, an author and teacher working in the 2nd century CE, Anaximander developed a sophisticated sundial, indicating solstices and equinoxes as well as seasons (Figure 3); this may have been done in collaboration with the architect Theodorus of Samos. Anaximander's familiarity with architectural techniques possibly stimulated other ideas. A Christian writer of the 3rd century CE, Hippolytus of Rome, who summarized the views of early Greek philosophers in his *Refutations of All Heresies*, tells us that Anaximander thought that the Earth is shaped like a column drum, with the inhabitants walking on the flat surface on one side. The analogy to the column drums of monumental buildings may have been inspired by new construction techniques used to create huge columned temples, such as the Didymaion, the temple of Apollo at Didyma, near Miletus (Figure 4). Analogies—especially to well-known objects and phenomena—were used in many explanations of nature, and not only by Anaximander.

Anaximander was also known for his written work, but only fragments of his *On Nature* survive. He may have created a new genre: the treatise on nature, characterized by both its subject matter and order of topics, beginning with the origin of heaven and Earth, and ending with human beings. Simplicius commented on his writing style as well as his ideas, indicating that he had read at least part of the work. Using information collected and preserved by Theophrastus, Simplicius reported that Anaximander named the *apeiron* ('the boundless' or 'the unlimited') as the principle and element (*stoicheion*) of everything that exists. Some have understood the *apeiron* as material stuff, but its character is puzzling.

3. Anaximander is sometimes thought to be the man holding the sundial on this 3rd-century CE Roman mosaic.

Anaximander suggested that the heavens and multiple worlds (*kosmoi*) come into existence from the *apeiron* and they also pass away. Whether the different worlds coexist simultaneously or exist only one after the other is not clear. Several sources report that Anaximander theorized that humans were first born inside fish, emerging and coming to land when old enough to fend for themselves. Because humans need to be nurtured over a long period, they would not have survived if they had begun in their current form. Simplicius provides what appears to be a quotation from Anaximander, stating that what comes to be and passes away does so 'according to necessity, for they pay penalty and restitution

4. Columns of the Temple of Apollo at Didyma, near Miletus, built in the Hellenistic period. The earlier 6th-century temple was destroyed by the Persians in 494 BCE.

to each other for their injustice, in accordance with the ordering of time'. For Simplicius, this seemingly moral (and politically inflected) message was somewhat poetic.

Others also shared information about Anaximander. According to Aëtius, the author of a survey of Greek natural philosophy and probably at work in the late 1st century CE, Anaximander said that thunder, lightning, thunderbolts, whirlwinds, and typhoons occur as a result of *pneuma*. The word *pneuma* has several meanings, including wind, breath, spirit, air, and even 'a blowing', that is, moving air. *Pneuma* also plays a role in some later medical contexts. Aëtius' survey may have been indirectly derived from the *Opinions of the Physicists* or *Physical Opinions* by Theophrastus, who also wrote a work *On Winds*, in which he described wind as air in motion; this seems to have been a topic of particular interest to Theophrastus, who may have been responsible for stressing the importance of moving air in Anaximander's account. This is just the sort of situation, with regards to our sources for early philosophers, that makes it difficult to know who thought what.

Anaximenes of Miletus

While it is unclear how important *pneuma* was to Anaximander, according to Aristotle his student Anaximenes of Miletus named air (*aer*) as the primary form of matter. But it is not certain how to understand this term *aer*; it may refer to a sort of mist. Simplicius, in his commentary on Aristotle's *Physics*, shared Theophrastus' explanation of Anaximenes' view that air undergoes changes and takes on other material forms as it condenses or becomes rarefied. In this way, air provides the basis of other matter, including clouds and wind. Aëtius reports that Anaximenes claimed that all things come to be from and dissolve back into air, which encloses the whole world. He apparently regarded air—as wind or breath—as fundamentally involved in the workings of the world in a way similar to that in which the soul, also composed of air, functions in humans. The idea that what happens in the world as a whole is similar to what happens in human beings is a recurring theme in ancient accounts of nature. Some such suggestions depend on an analogy between the cosmos and humans, while others point to a relationship, in which the larger (macro-) world is mirrored in a miniature (micro-) version—a macrocosm and a microcosm.

Parmenides of Elea

Knowing what we can and cannot know was (and still is) a fundamental question for many philosophers, scientists, and others. While some of the early philosophers focused on the material composition of nature, Parmenides of Elea was preoccupied with the question, 'How do we know?' Fragments of his writing—a poem in epic hexameter—were preserved by Sextus Empiricus, probably around 200 CE, and by Simplicius, in the 6th century CE. Parmenides describes how he began his intellectual journey, driven in a chariot drawn by mares and led by the daughters of the Sun, Helios, to the gates of the paths of Night and Day, overseen by Justice. He is welcomed by a goddess, who plays a major role in what follows. Why did he choose poetry to present his philosophical views? By echoing the style of the Homeric and Hesiodic poems, Parmenides located himself within a powerful tradition. However, it is not entirely clear whether he was attempting to enlist their authority or subvert it, offering a very different account of the world.

Parmenides explained that he was following the goddess's directions. The idea of divine inspiration—of a god or goddess acting as a muse—was a powerful one. Later philosophical authors and poets similarly pointed to divine influence. Parmenides' goddess plays a key role, showing him the Way of Truth and then the Way of Opinion (*doxa*), describing what is, in contrast to what is not. We can have knowledge only about what is, and cannot gain true understanding of what is not. What exactly Parmenides meant by this has been the subject of much debate. Scholars agree that he was preoccupied with ontology and epistemology, subjects that continue to captivate philosophers. Parmenides' Way of Opinion included much of what we regard as 'scientific'. In antiquity, he was credited with having provided descriptions and explanations of the natural world, including the idea that the light of the Moon comes from the Sun, and the identification of the morning star and the evening star as the same.

Heraclitus of Ephesus

Many *physiologoi* offered explanations of astronomical phenomena. Heraclitus of Ephesus, who flourished around 500 BCE, was credited by Diogenes Laertius with having written a work, *On Nature*. He described the astronomical bodies as bowls, with their hollow interiors facing us. Within these, exhalations collect, forming flames. The Sun is the brightest and hottest, and the Moon's bowl gradually turns, resulting in the lunar phases. When the bowls are turned upwards, the Sun and Moon are eclipsed. This vivid visual description explains astronomical phenomena in a way that is readily comprehensible, invoking everyday objects—bowls—that we can easily imagine in the sky.

Heraclitus is famous for saying that we cannot step into the same river twice. Here, something familiar—a river—serves as the exemplar of the ever-changing, something that may be much harder to contemplate. Different versions of what Heraclitus was said to have claimed survive, and people debate what he actually meant. He seems not to have been speaking about rivers or water per se, but rather making a more general point that change is continuously taking place. The river changes, while remaining the same. For Heraclitus, there is a unity of opposites, as opposites coincide. Even in antiquity Heraclitus had a reputation for being difficult to fathom: some refer to him as 'the obscure'.

Heraclitus was reported to have said that neither gods nor men made the cosmos. It has no beginning nor end, but is a fire, burning and extinguishing equally, so that it continually blazes. This rich image is similar to that of the river that is never quite the same, but nonetheless always flowing. For Heraclitus, the world was always changing, yet its structure—the world order, the cosmos—remains the same.

Empedocles of Acragas

Powerful imagery was useful for explaining nature, enabling others to picture what was being described. Empedocles, who was active in the 5th century BCE and from the Greek city of Acragas in Sicily, is credited with two poems, one referred to as *Purifications*, the other *On Nature*. Aristotle, in his *Poetics*, made clear that he didn't think much of Empedocles as a poet. For him, Homer and Empedocles shared nothing other than metre, so that while the former could be rightly called a poet, the latter should be regarded as a physicist. But others disagreed;

Plutarch (born before 50 CE–died after 120) argued that Empedocles was not trying to be showy with his use of language, but employed simple descriptions to convey essential information, such as 'cloud-gatherer' (for the air), and 'rich in blood' (of the liver). These epithets are reminiscent of the Homeric poems (think of the 'wine-dark seas'), conjuring up graphic images.

Whatever Aristotle thought of Empedocles as a poet, he valued his ideas about nature. He credited Empedocles as the first to identify four elements (referred to as 'roots' by Empedocles): Fire, Air, Water, and Earth. Aëtius identified Empedocles' roots with gods and goddesses—Zeus, Hera, Nestis, and Aidoneus—but it is not clear which divinity was associated with which specific element.

Empedocles' interests and skills were actively practical as well as philosophical. Diogenes Laertius reports that he wrote a work on medicine as well as other poems, tragedies, and political works. He was apparently a physician and a skilled orator, and admired for devising a way to limit crop damage by stretching out bags made of asses' skins to catch gusts of wind. For this innovation, he was called 'Empedocles halting the winds' or 'wind-stayer', an epithet strikingly similar to those given to the traditional gods.

Anaxagoras of Clazomenae

Understandably, success in dealing with threatening weather attracted admiration. Another early philosopher, Anaxagoras of Clazomenae, was reported to have travelled to Olympia wearing a leather cloak, as if expecting a downpour. It did then rain. The way the story is told, we sense that he had not simply covered up 'just in case'. But we are not told how he managed to make an accurate prediction, or how often. Rain also featured in Anaxagoras' explanation of rivers, which—according to Hippolytus—he said come from rain and water contained in the hollows of the Earth. He suggested that the rising of the Nile in summer is caused by water from melting snow flowing into it from southern regions. Hippolytus reports that Anaxagoras said that animals originally arose in moisture, but afterwards from one another.

Like many others of the *physiologoi*, we have reports not only of Anaxagoras' ideas about what happens on Earth, but also in the sky. According to Plutarch, Anaxagoras thought the Moon was as large as the Peloponnesus. Hippolytus states that Anaxagoras thought the Sun was larger than the Peloponnesus and that the heavenly bodies were fiery stones, with some carried around with the Sun and Moon but invisible to us. The heat of the stars is not felt by us because they are so far away, and they are not as hot as the Sun, as they are in a colder region. The Moon is eclipsed when blocked by the Earth or one of the other bodies below it; the Sun is eclipsed when the Moon blocks it, during the new moon phase. Similarly to Parmenides, he thought that the Moon received its brightness from the Sun; he also explained that rainbows are the reflection of the Sun in clouds. Anaxagoras reputedly stated that the Earth is in the air, the air itself being powerful enough to hold it in place, and that the Moon is somewhat similar to the Earth, with plains, mountains, and valleys.

Anaxagoras is also credited by some modern scholars, as well as ancient, with having used diagrams. Aristotle praised him for appearing as a sane person after the random chattering of those who preceded him, especially for asserting that intelligence exists in nature, as it does in animals, and that this is responsible for the world order. According to Diogenes Laertius, Anaxagoras said that in the beginning, all things were together, then Mind (*nous*) came and set them in order. For this reason, his nickname was 'Nous'. Simplicius, our principal source for Anaxagoras' fragments, reports that he said that in everything there is a portion of everything, except Mind, which is unlimited (*apeiron*), self-ruling, mixed with nothing and alone.

Leucippus and Democritus

During the 5th century BCE, ideas about what things are made of took a different turn with Leucippus of Miletus and Democritus of Abdera. They are reputed to have thought that things are composed of small pieces of matter that cannot be divided or cut. Our word 'atom' comes from the Greek *atomos*, meaning 'uncuttable', and Leucippus and Democritus are known as 'atomists'. Aristotle is an important source for our knowledge of atomism, but he did not distinguish between the views of Leucippus and Democritus. No complete writings by either survive. However, ancient reports credit each of them with ideas on other topics, including cosmogony, cosmology, astronomy, and meteorology, and they are not always reported to have shared identical views on all things. For example, Aëtius tells us that Leucippus thought the Earth was shaped like a drum, while Democritus thought it disc-shaped, and concave in the middle. Leucippus is reported by Diogenes Laertius to have regarded the heavenly bodies as fiery. Democritus is one of the authorities cited in later Greek astrometeorological calendars.

Pliny the Elder credits Democritus with having been the first to point to the alliance (*societas*) between the heavens and the Earth. Whether or not the story is true, he also reports that Democritus' wealthy fellow-citizens disparaged his studies of nature. To make a point, Democritus adopted a similar tactic to that allegedly used by Thales, employing astronomical knowledge to predict a rise in the price of olive oil. He bought up all of the available supplies, accrued great wealth, and then returned the money he had made, as he merely wanted to show that he could become rich if he wished. Like so many of the early Greek philosophers, he was renowned for his practical know-how as well as his theoretical ideas.

Chapter 4
Those clever Greeks

In the 5th century BCE, some Greek authors referred to ideas about nature first expressed by the *physiologoi*, even when their own works were not primarily focused on explaining nature. A number of these writers comment upon—and even ridicule—philosophers and their attempts at explanation. Their works give a sense of the wider circulation of such ideas, and evidence of how attempts at explaining nature were regarded.

There are indications that various people—including some of those attending Athenian plays—were acquainted with the ideas of those who had reputations for philosophizing about nature. Diogenes Laertius reports that the 5th-century BCE Athenian playwright Euripides, who had studied with Anaxagoras, called the Sun a 'golden clod' of earth in his (lost) play *Phaëthon*, presumably adopting an idea from his teacher. The Roman author Seneca, in his *Natural Questions*, suggests that other dramatists, including Sophocles and possibly Aeschylus, shared Anaxagoras' views about the source of the Nile, and that Aristophanes, the author of comedies, was also familiar with his ideas. Anaxagoras' thoughts on nature would probably have been known by at least some members of the audience, comprised of Athens' citizens, for whom theatre was central to civic and cultural life. While theatre spectators possibly did not include women, they did include free Athenian males, entitled to vote.

Aristophanes

Aristophanes, who was active during the late 5th and early 4th centuries BCE, is a surprisingly rich source regarding contemporary scientific endeavours. In *Birds*, the character Meton describes how he uses a ruler to inscribe a square within a circle, drawing streets so that they converge in the centre at the marketplace. Not so impressed, Pisthetaerus jokingly calls him 'another Thales!' And in *Clouds*, Aristophanes pokes fun at those who discussed nature, caricatured by someone named 'Socrates'. The 5th-century BCE Diogenes of Apollonia is often assumed to have served as a model for some of the speeches delivered by this 'Socrates'. He may have been a physician, originally from the Milesian colony Apollonia, who spent time in Athens, as Athenian philosophers mention him by name. We know about his theory of *phlebes* (blood vessels or vascular channels) from Aristotle, and Theophrastus (quoted by Simplicius) describes him as an eclectic writer, whose ideas are linked to those of a number of early philosophers. His notion of the primacy of air may be related to that of Anaximenes.

One of the comedic speeches—delivered by 'Socrates' while suspended in a basket in the air—gives a sense of Aristophanes' own view of the nonsensical flavour of some of the study of nature:

> I'd never be able
> To investigate all higher matters correctly
> Without elevating my intellect and thought
> And mixing my delicate mind with the kindred air.
> If I studied the things above from down on the ground
> I would never have made discoveries, since the earth
> Draws down by force the moisture of our thoughts.
> The phenomenon's just the same with watercress.

Aristophanes gives us the strong impression that those who investigate nature are airheads. Diogenes of Apollonia was not the only target of his barbs, for if scientific theorizing had been limited to only a few people, the point of the parody would have been lost on his audience. By using comedy to criticize the ideas of those he implied were always talking about clouds and the like, Aristophanes provides us with amusing evidence of the reactions of some ancient Greeks to others who sought to explain natural phenomena.

Herodotus

In addition to the Athenian playwrights, other 5th-century BCE authors flaunted their scientific knowledge. Herodotus is often regarded as the Father of History, and his *Histories* may have been the single longest prose work produced in Greek before the 4th century BCE, providing an account of the wars between the Persian Empire and Greek city-states during the first part of the 5th century. But that's not all he discussed. He offers details of people living in various places, with different customs and points of view. And he may also have been the Father of History of Science, as he reports scientific ideas, discoveries, and inventions.

Not everyone who encountered Herodotus' *Histories* would have read it themselves: reading was often done aloud, sometimes for an audience. Herodotus himself may have performed public readings of some parts of his work. The engaging tone of the *Histories*, which includes many anecdotes and tantalizing details of other places and peoples, may have been designed to attract a broad audience.

In the opening section, Herodotus refers to his work as *historia* (enquiry). His aim was to identify causes, primarily the causes of wars. But he was clearly interested in explaining many other things too. He notes that Greeks wished to be known for their cleverness, and that one of the ways that they demonstrated this cleverness was by offering accounts of natural phenomena. He almost suggests that this was something of a national pastime, and displayed his own fascination with such explanations.

Curious about the seasonal flooding of the river Nile, Herodotus complains that he had failed to extract any information from Egyptian priests, or ordinary Egyptians:

> I was particularly eager to find out from them why the Nile starts coming down in a flood at the summer solstice and continues flooding for a hundred days, but when the hundred days are over the water starts to recede and decrease in volume, with the result that it remains low for the whole winter, until the summer solstice comes round again.... I asked them what it was about the Nile that made it behave in the opposite way from all other rivers.

Herodotus remarks that while no one in Egypt could give a reason, three separate theories had been put forward by different Greek thinkers, motivated by the desire to enhance their own reputation for being clever. Herodotus singles out Greeks as being the only ones interested in elucidating the Nile's flooding.

In discussing the three explanations, Herodotus almost immediately discounts two of them. One of these suggests that the north winds in summer cause water to rise by retarding the current's flow towards the sea. He counters this by noting that on many occasions these winds have failed to blow, and the Nile has nevertheless risen as usual. He also suggests that if in fact these winds were responsible for the Nile's rise, other rivers would certainly be affected in much the same way. He then points to such rivers in Syria and Libya that do not behave in that way. Herodotus rejects this explanation because it does not fit with observation and experience.

He also objects to the second explanation, that the Nile behaves in the way it does because it flows from Ocean, the circumglobal stream described by Homer. Here, Herodotus' objection is that this account seems to lack any factual basis. He complains that 'I do not know of the existence of any River Ocean, and I think that Homer or one of the other poets from past times invented the name and introduced it into his poetry.' The authority of the poet is not sufficient to endorse his views on nature, which Herodotus regarded as somewhat fantastical.

The final theory offered by the Greeks is that the Nile's water comes from melting snow; this was the view held by Anaxagoras, but Herodotus does not mention any names. According to Herodotus, this theory is more plausible than the others, and yet also furthest from the truth, because the Nile flows from Libya through Ethiopia into Egypt, from a very hot climate to a cooler one. Since this is the case, how could the Nile possibly originate from snow?

After rejecting theories of (other) Greeks, Herodotus offers his own detailed explanation. He seems to be making a statement that one need not be a philosopher to be able to offer an excellent and rational account of natural phenomena. According to him, the position of the Sun, as an important heat source, is key. Because the Nile is close to the course of the Sun, it is affected more by the Sun's motions than other rivers and behaves in a way that is completely different. Herodotus' original question about the Nile's behaviour linked it to astronomical events (namely, the summer solstice), and his own explanation of the Nile's flooding crucially involves not only the river itself, but also the Sun. (Today, we understand the seasonal rain in the Ethiopian highlands as responsible for the annual summer flood of the Nile.)

Herodotus was from what is now western Anatolia, the region that was also home to Thales and the other Milesians; he was presumably aware of their philosophizing. He uses what would today be called an evidence-based approach to account for phenomena (at least the flooding of the Nile river). By describing his work as *historiai* ('enquiries'), Herodotus was signalling that his work was part of a broader world of enquiry, not only into conflict and human nature, but about nature itself. While we do not normally think of Herodotus as a scientist, he thought that his own explanation of the Nile's flooding was better than those offered by others who had a reputation for cleverness in understanding nature.

Socrates

Socrates—the real Socrates—was roughly a contemporary of Herodotus, but somewhat younger. We know him primarily through his depiction in the dialogues of his student Plato. In the *Phaedo*, Plato presented him as recalling how as a young man he had a strong desire to understand nature. There, Socrates explains his initial ambition:

> it seemed to me splendid to know the reasons for each thing, why each thing comes to be, why it perishes, and why it exists. And I was always shifting back and forth, examining, for a start, questions like these: is it, as some said, whenever the hot and the cold give rise to putrefaction, that living creatures develop? And is it blood that we think with, or air, or fire? Or is it none of those, but the brain that provides the senses of hearing and seeing and smelling, from which memory and judgement, when they've acquired stability, that knowledge comes to be accordingly? Next, when I went on to examine the destruction of those things, and what happens in the heavens and the earth, I finally judged myself to have absolutely no gift for that kind of inquiry.

Plato's Socrates vividly lists the sorts of enquiries that were part of the *historia* about nature. And he expresses disappointment that in his day natural science focused on identifying the material causes of things rather than the reason they exist. Teleology—understanding the *telos* or purpose of things—became a central concern for many who followed Socrates, including those who sought to explain the world using natural philosophy and mathematics.

Chapter 5
Let no one unskilled in geometry enter

Not all of the early philosophers thought that matter was fundamental to understanding nature. The idea that the structure of the cosmos was based on numbers—rather than some material such as water or air—is credited to Pythagoras and his followers. Although he apparently wrote nothing himself, Pythagoras is famous, and not only because of the theorem that today bears his name. Diogenes Laertius regarded Pythagoras as the originator of philosophy in Italy. He reports that a certain Apollodorus, who was knowledgeable about calculations, claimed that Pythagoras had made a special sacrifice of oxen (a hecatomb) on discovering (and this is the word used: *euronta*) that on a right-angled triangle the square of the hypotenuse is equal to the sum of the squares of the two other sides. He marked his mathematical discovery with a religious ritual, a sacrifice. Pythagoras himself appears to have had cult status, for his followers used his teachings—the 'things heard', passed down orally—as guidelines for a way of life.

The so-called Pythagoreans

It is notoriously tricky to tell whether the ideas we think of as 'Pythagorean' were actually Pythagoras' own. Much of what we (think we) know about Pythagoras and those whom Aristotle referred to as 'the so-called Pythagoreans' was passed down, and probably filtered, by Plato and members of his school, who used Pythagoras and other early philosophers to construct an intellectual history for—and so bolster and consolidate—their own work. In several dialogues, including the *Timaeus*, Plato highlights the Pythagorean idea of the mathematical structure of the cosmos, and the place of humans within that mathematically determined universe.

The original concept of the harmony of the spheres is usually traced to early Pythagoreans. The Greek word *harmonia* refers to a 'means of joining or fastening'. In Homer's *Odyssey* a form of the word refers to the joining together of the boards of Odysseus' raft. The original sense of *harmonia* was extended to refer to a framework like the human form, an agreement or covenant, order, and the method of stringing musical instruments to fit a musical mode or scale. Diogenes Laertius reports the views of Philolaus, living in the 5th century BCE and originating from what is now southern Italy, who claimed that all of nature was joined together, that is, harmonized: the whole world and all the things in it. Timaeus, the main speaker in Plato's dialogue of that name and understood to be a Pythagorean, explains that hearing was given by the gods not only for pleasure but to allow perception of the cosmic harmony, through which humans may then similarly order their own souls. As in Anaximenes' ideas, the human soul itself shares something with the world order.

Aristotle argues against the idea that the astronomical motions produce any sound. He explains that advocates of the view believed that large bodies must inevitably produce sound while moving and, furthermore, that the speed of the motions of the heavenly bodies and their distances are in the same mathematical relationships as concordant musical sounds. He thought the idea was elegant, but ultimately rejected it, claiming that such a loud noise would surely be as destructive as thunder, which can cleave apart rocks. He rejects the possibility of any sound being associated with the celestial motions, drawing an analogy to the quiet way in which ships travel downstream. But not everyone was willing to give up the idea of celestial harmony and, much later, Pliny the Elder admits that he was himself unable to choose between the two views.

For Philolaus, the world order included fire at the centre, rather than the Earth. The association of Pythagoreanism with a non-geocentric universe continued into the early modern period, when in 1543 Nicolas Copernicus cited Philolaus as his precursor, in *De revolutionibus orbium cœlestium* (*On the Revolutions of the Heavenly Spheres*) (Figure 5). Another important feature of the Pythagorean heritage is the idea—especially influential for mathematics and science—that there are four kindred mathematical sciences: arithmetic, geometry, harmonics, and astronomy, sometimes referred to as the Pythagorean sisters. The conviction that the order and structure of the world are based on mathematics has exerted great influence over thousands of years, and Plato himself was a powerful promoter of this concept.

Plato and the fundamental importance of mathematics

Plato was the founder of the Academy in Athens, which was said to have had an inscription over the door stating, 'Let no one unskilled in geometry enter'. Nevertheless, we know from Plato's *Republic* that future philosopher-kings were required to study several types of mathematics, not only geometry. Their education was meant to begin with training in music and gymnastics, followed by a course of studies including arithmetic, geometry, stereometry, astronomy, and harmonics. These mathematical studies were preliminary to that of dialectic, vital to the search for truth. Mathematics is not only important for the education of philosopher-kings: it is crucial to the underlying structure of the world.

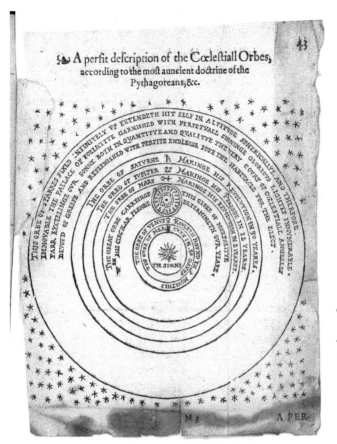

Let no one unskilled in geometry enter

5. Image from 'A Perfit Description of the Cœlestiall Orbes' by Thomas Digges, appended to the 1576 London edition of *A Prognostication Everlasting*, by his father, Leonard Digges. Thomas Digges was the first author in England to publicly advocate Copernican cosmology.

Athens became a major centre for philosophy in the 5th and 4th centuries BCE. The philosophical schools founded there promoted scientific work of all kinds, including philosophy about nature. Plato's ideas and those of other members of the Academy have had a lasting impact on science. Of particular influence was their concentration on the fundamental epistemological query: how do we obtain—and decide what counts as—knowledge? According to Plato, the philosopher strives for knowledge of true reality, which cannot be seen and can only be apprehended by thought. We do not receive knowledge through our senses. Our souls are able to recollect knowledge of the eternal unchanging Forms (or Ideas) gained before the soul entered our bodies. When we do acquire knowledge, we are using that part of our soul whose nature it is to be wise.

Plato emphasized the value of being educated in mathematics; astronomy was one of the key branches of mathematical study. In the *Republic*, Socrates describes how best to pursue astronomy: by means of problems, as in the study of geometry. We are told to let the things in the heavens alone, if we are to have true astronomy. The instructions for undertaking astronomy are offered in the context of Socrates' discussion of the education necessary for future leaders, not necessarily for those focused on the natural world. Astronomy is presented as a stepping-stone towards more fundamental enquiry. Until well into the 17th century, if not later, mathematics was understood to be the language of astronomy.

Plato is credited by one tradition with having set the fundamental problem of classical astronomy: to use uniform circular motions to 'save' astronomical phenomena and account for the apparent motions of the planets, the 'wandering' stars (which include the Sun and Moon, as well as Mercury, Venus, Mars, Jupiter, and Saturn) that seem to move relative to the 'fixed' stars. The requirement to 'save the appearances' is taken to mean that we must—at least to some extent—account for what we see in the sky. This seems to contradict Socrates' instructions that we should leave the things in the heavens alone and proceed by means of problems as in geometry, using deduction rather than experience. This apparent contradiction highlights an important tension in astronomy and other sciences in antiquity as well as the present day: to what extent are we seeking an elegant theory, and to what extent must that theory accord with phenomena? We are not certain that this tradition interpreted Plato's views as he intended. Nevertheless, the use of uniform, circular motion to explain the apparent movements of astronomical bodies was a key feature of Greek astronomy, so much so that when Johannes Kepler, in the early 17th century, adopted ellipses rather than circles to describe astronomical motions, he recognized that he had violated an important tradition.

Astronomy's reliance on geometrical accounts of the motions of astronomical bodies distinguished it from cosmology and the explanation of the order of the cosmos. In contrast to the Presocratics, whose writings do not survive in their entirety, Plato's dialogues are extant. However, partly because of the dialogue form, with different interlocutors sharing their views and

Plato's influence

For much of the Middle Ages in the Latin West, if you referred to Plato you were probably thinking of the *Timaeus*. It answered crucial questions about the world and was the only Platonic dialogue continuously available in Latin translation. Prior to the Renaissance, two Latin versions, both incomplete, were in circulation: Cicero's, from the 1st century BCE, and Calcidius', from the 4th century CE. As a creation story emphasizing the cosmic significance of mathematics, the *Timaeus* offered evidence that the creator was a mathematician at heart. Johannes Kepler regarded the *Timaeus* as a Pythagorean interpretation of the first chapter of the biblical book of Genesis. He approvingly wrote to Galileo Galilei (1564–1642), congratulating him for 'following the lead of Plato and Pythagoras, our true masters' in emphasizing the importance of mathematics.

discussing ideas, we don't always know what Plato himself thought. The dialogues can be read as works of literature as well as philosophy. Timaeus presents a powerful idea, with a distinguished legacy, that mathematics informs the structure of the world. However, we do not know precisely to what extent his account mirrors Plato's own ideas on cosmology and cosmogony.

The mythic and the scientific, together

Timaeus is lauded as having expert knowledge in Plato's dialogue of the same name. Another speaker, Critias, explains that as Timaeus 'knows more than the rest of us about the heavenly bodies and has specialized in natural science, we decided that he should speak first, and should start with the origin of the universe and end with the creation of human beings'. As Timaeus begins, he warns that his audience should not be surprised if 'when discussing gods and the creation of the universe, we often find it impossible to give accounts that are altogether internally consistent in every respect and perfectly precise', arguing that

> we'll have to be content if we come up with statements that are as plausible as anyone else's, and we should bear in mind the fact that I and all of you, the speaker and his judges, are no more than human, which means that on these matters we ought to accept the likely account and not demand more than that.

Timaeus presents his account of the origin of the world as a 'likely story', with many seemingly mythical elements. He rejects the possibility of knowing the truth: any attempt to search for a completely certain account of the world would be futile, and human beings can only hope to offer a plausible explanation.

Both the subject matter and order of topics are characteristic of the genre of treatises on nature inaugurated by Anaximander, beginning with the origin of heaven and ending with humans. Timaeus' speech can also be read as a creation myth, in which the

Demiurge, a craftsman of sorts, fashions the world, its contents, and its inhabitants. The use of myth enables Plato to convey 'truth' in a manner that is not necessarily intended to be taken literally. (In another Platonic dialogue, the *Protagoras*, the main interlocutor holds that the truth of things can be conveyed in different ways, via ordinary discourse (*logos*) or alternatively through *mythos*, the mythic.) The use of myth as a tool for explanation has a degree of built-in indeterminacy that is useful for Plato's purpose. This enables Timaeus to emphasize the provisional character of his account, appropriate—in Plato's view—for an account of the physical world apprehended through our senses, and which therefore, according to his theory of knowledge, cannot be known in the same way as intelligible being, apprehended through intellect alone. The creation account in the *Timaeus* includes numerous shifts between what reads as myth, fable, scientific analysis, and philosophical argument, and it would be difficult and misleading to suggest that there is a clear-cut division between the different modes of explanation.

Readers since antiquity have sought to understand the passage, regarded as both astronomical and cosmological, in which Timaeus describes the fashioning of the celestial sphere by the Demiurge, employing mathematical language that refers to arithmetic (using numbers) and geometry (involving circles). Timaeus offers a vivid description of the way in which the Demiurge constructed circles in the World-Soul, the soul of the entire universe, using mathematical proportions. The motion of the circle of the Same is considered to be a motion of the World-Soul, manifested in the sphere of the fixed stars; the circle of the Different is associated with the motions of the planets, including the Sun. Some readers would have been mystified by Timaeus' terminology, including words such as 'diagonal'. Centuries later, Plutarch, who was very familiar with Plato's writings, referred to young men showing off their mathematical knowledge by speaking about the meaning of 'by way of the diagonal'.

Timaeus claimed that it is impossible to describe the motions of the planets without visible models. Some commentators have suggested that Timaeus was describing an astronomical device, a sort of armillary sphere, used for teaching (Figure 6).

According to Timaeus, because the cosmos is a whole whose parts fit together in a harmonious way, we, as beings created as part of that universe, can attain the best life possible by gaining knowledge of the cosmic harmonies. *Harmonia* is used in at least two related ways, referring to the cosmic fitting together and also to the divine sound of music that can be heard by mortals, allowing them to imitate the cosmic harmony. According to Plato, our senses of hearing and sight bring us into contact with these

6. **Armillary sphere, 15th century (on a later base).**

harmonies. The *Republic* emphasizes the value of our senses in comprehending the mathematical structure of the cosmos. As the eyes are made for astronomy and the ears presumably for harmonics, according to the Pythagoreans, these are related branches of knowledge.

Turning away from the traditional gods and their perceived lack of morality, Plato emphasized visible gods, demonstrating purposeful order in the world and invoking an ethics based on the emulation of that order. In the *Timaeus*, the heavenly bodies are gods. Furthermore, each human soul had a portion of the immortal soul created by the Demiurge. According to Plato, of the various human sensations only sight and sound touch the soul. The visibility of the celestial bodies was part of the cosmic plan; the Demiurge made them mostly of fire, so that they would be bright and visible. Vision is the sensation of motion and allows the orderly motions of the heavens to be carried to our souls. Unlike the traditional gods of myth who reveal themselves only so far as they wish, the celestial bodies are readily seen by all. This visibility was purposefully ordained by the Demiurge, who, when he devised the orderly motions of the visible universe, provided a means by which this order could be recognized. Once the ordered motions of the cosmos have been perceived, every human has several ways by which to imitate that order. However, according to Timaeus, only those who are philosophers will be able to obtain the fullest measure of immortality available to humans.

Elsewhere Plato suggested that all citizens—even those not training as philosophers—should undertake a certain amount of astronomical study, not only for its practical value but for religious reasons, to learn enough about the heavenly gods to prevent blasphemy of them, and to ensure piety in sacrifice and prayer. For Plato, it is an ethical necessity for all to be familiar with the celestial bodies and their motions. Even if one is not capable of attaining true knowledge of the heavens, one should at least acquire a correct opinion. The religious value accorded the study of astronomy is echoed in the *Epinomis*, which was probably written by a member of the Academy, if not Plato himself. There we learn that the greatest of human virtues is piety, which can be learned by studying astronomy, for it offers understanding of the creation of divine things, the most beautiful things that humans can see.

As part of his account of the origin of the world, Timaeus describes the creation of men, women, and those animals that are not human. The Demiurge himself made the immortal part of individual souls, delegating the creation of human bodies to lesser, created gods. In Plato's mathematical universe, shape matters: the human head is fashioned as a sphere, copying the shape of the cosmos. The rest of our bodies—our limbs—are there to serve as a vehicle to transport the head, which holds the immortal, rational part of the soul during our lifetime.

Archytas of Tarentum

Plato was not alone in tackling questions about the cosmos and the motions of the astronomical bodies. Archytas of Tarentum, a noted Pythagorean who knew Plato, is credited with one of the oldest thought experiments in the philosophical literature, regarding spatial infinity. Versions of it are discussed by later thinkers as diverse as Lucretius, John Locke, and Isaac Newton. Our source—once again, Simplicius, in this instance quoting Eudemus of Rhodes, a student of Aristotle—tells us that Archytas asked whether, if he came to the edge of the heavens, he would be able to reach out his hand or a stick. Archytas argued that to not be able to reach out would be absurd, but if he were able to then what was outside must be either a body or a place. This action could be carried on infinitely, always reaching beyond the previous limit. Not everyone accepted this argument. For at least one of Plato's students, Aristotle, the cosmos was spatially finite and strictly bounded, with nothing existing beyond that boundary.

Eudoxus of Cnidos

Eudoxus of Cnidos may have been a student of Archytas, in the 4th century BCE. Very little about his life and work is known with any certainty. Diogenes Laertius reports that he was called 'Endoxus' ('illustrious' or 'reputable'), in a play on words, because of his brilliant reputation. None of his writings survive, only accounts and fragments in other ancient authors. There is evidence that he founded a school at Cyzicus, in the Hellespont, which may have survived into the 3rd century BCE. Anecdotes relate his invention of a type of sundial, and his introduction of the practice of arranging furniture in a semicircle to accommodate more people (Figure 7).

Today Eudoxus is primarily known for his geometrical model of planetary motion. According to an ancient account, he was the first to answer Plato's challenge to account for the apparently irregular movements of the astronomical bodies. Aristotle relates that he devised a system of nested homocentric spheres to describe the apparent motions of each of the heavenly bodies. Writing much later, Simplicius explained that according to Eudoxus the path of a planet describes a figure called a 'hippopede' or horse-fetter (looking a little like a modern numeral 8 on its side). Eudoxus was able to account for various phenomena and 'save the appearances' in an approximate way, although not everything could be explained by his model, such as the changing brightness of planets.

Eudoxus produced a prose work in Greek, known as the *Phainomena*, in which he described the constellations and provided information about stellar risings and settings. This work was adapted by Aratus, who composed a poem, also in Greek and with the same name, which included information on weather signs in addition to astronomical phenomena. Eudoxus' followers in the school of mathematicians at Cyzicus seem to have had a

7. **Instruments, including a sundial and globe, used by men meeting in a garden, depicted on a Roman mosaic, 1st century BCE–1st century CE.**

special interest in building instruments to demonstrate his ideas of planetary motion. Aristotle adds that Callippus of Cyzicus, working in the 4th century BCE, adopted Eudoxus' approach and increased the number of spheres, in order to better explain the phenomena. Aristotle objected to the number of spheres suggested by both Eudoxus and Callippus, offering his own improvements to their approach and further increasing the number.

The notion of 'saving the phenomena'—offering a solution that can account for observations even if not explaining the cause of those

Astronomical poetry

In antiquity, Aratus' poem enjoyed extraordinary success, inspiring at least 27 commentaries, including one produced in the 2nd century BCE by the noted astronomer Hipparchus, and a number of translations into Latin including those by Cicero and an author presumed to be the general and heir-designate of the Roman Empire, Germanicus Julius Caesar, writing in the early 1st century CE. Each had their reasons for producing a new version, including—in the case of Germanicus—the updating of scientific data. The translations and reworkings of Aratus' poem demonstrate its enduring interest, overshadowing the work by Eudoxus upon which it was based. Eudoxus' text did not survive, whereas we know of at least six ancient Latin translations of Aratus' work. The Roman poet Marcus Manilius composed his own account of astronomy in Latin verse, the *Astronomica*, conveying detailed technical knowledge. In 1675, Edward Sherburne published an English translation of the first book of Manilius' work, promoting modern astronomy by depicting practitioners as participants in an august tradition dating back to antiquity (Figure 8).

phenomena—is associated with Plato and his followers. Questions regarding the application of mathematical methods to the explanation of natural phenomena are of great interest to historians and philosophers of science, and Eudoxus' work often figures importantly in their discussions. Pierre Duhem (1861–1916), a physicist, philosopher, and historian of science keenly interested in the relationship between mathematics and physics, was a proponent of an instrumentalist view of science, according to which theories permit predictions that correspond to observations, but those theories do not necessarily describe physical reality. Many, including Duhem, have assumed that Eudoxus did not believe that his geometrical model had any

THE SPHERE
of
M MANILIVS
made
An English
POEM
by
Edward Sherburne. Efq;

C&LIQVE VIAS ET SIDERA MONSTRAT

NATVRÆ VNIVERSITAS.   VNIVERSITATIS INTERPRES

8. Title page of Edward Sherburne's 1675 English translation of the
first book of Manilius' poem.

physical reality. However, the paucity of our knowledge of Eudoxus' work makes it impossible to know the intention underlying his astronomical solution. Nevertheless, the emphasis on the use of geometrical models to explain celestial motions inaugurated an important phase in astronomy, which would far outlive the ancient Greeks.

Let no one unskilled in geometry enter

Chapter 6
A theory of everything

Plato's most famous student was Aristotle. He claimed that all men, by nature, desire to know, and it could be argued that Aristotle himself desired to know it all. Having been a member of Plato's Academy, he founded his own school also in Athens, the Lyceum, and produced—with the help of students and colleagues—systematic works on a broad range of subjects, including cosmology, physics, and meteorology, as well as detailed studies of animals. These scientific works were only a portion of his vast output, which covered politics, rhetoric, and literary theory. Aristotle and his colleagues carried out their own first-hand observations on a range of natural objects: Aristotle studied animals; Theophrastus, plants; and Aristoxenus, harmonics. Aristoxenus had previously studied with Pythagoreans, just as Aristotle had studied with Plato—a reminder that people did sometimes move between philosophical schools.

Aristotle had strong views about how to do philosophy, including philosophy of nature. He discusses different ways in which the term *physis* may be understood, sometimes referring to views of his predecessors. By one account, he says, nature is a thing's matter; by another, it is the shape or form specified in the definition of a thing. For Aristotle, nature is a principle of change. He prioritizes the study of change, arguing that if we don't understand that then we cannot have knowledge of nature.

The body (*corpus*) of Aristotle's work

Diogenes Laertius lists over 150 titles of works credited to Aristotle. Of these—if indeed they all ever existed—only a relatively small number survive. The Aristotelian Corpus is the collection of writings associated with Aristotle, preserved through transmission during the ancient and medieval periods. They are often described as philosophical and scientific treatises, but many of the works in the corpus are thought to be 'lecture notes' rather than polished pieces of writing intended for wider circulation. Nevertheless, many scholars believe that the wording of much of the material belongs to Aristotle. However, we don't know whether the grouping of 'books' (which may originally have been individual papyrus rolls) into a single work, or the ordering of those books, can be attributed to Aristotle. The titles and many of the introductory and concluding remarks may be due to later editors. Some works are referred to as 'pseudo-Aristotelian'; in those cases his authorship is doubtful and the actual author may be unknown.

Furthermore, for Aristotle, identifying causes is fundamental to gaining knowledge: in all things nature is the cause of order. This identification of 'nature' itself as a cause provides a powerful theme throughout his work and highlights the importance of finding causes when seeking to understand and explain.

Aristotle was a great collector of all sorts of things, including the opinions of others. He encouraged people to go through written handbooks of arguments, and draw up lists of different subjects, organizing them under separate headings. He recommended noting down the opinions of different thinkers. Lists or tables of diverse ideas on the same subject could be kept together, allowing the material to be found more easily when wanted. We can almost imagine something like an early form of a database, storing information for retrieval. You could ask a question, for example,

'What do people think about elements?', and find that Empedocles said that there were four. These lists and notes could then be used as a starting point in providing an argument in favour of a particular point of view.

Knowledge and explanation

Aristotle identified three kinds of rational knowledge (*epistēmē*): the practical, the productive, and the theoretical. This last kind is further divided into three: mathematical, physical, and theological knowledge. He discusses these and their relationships in several of his works, emphasizing different points. Whether or not an object of study is material and movable determines which type of theoretical knowledge is relevant. Natural (physical) science deals with material things that can move. For this reason, the study of motion is fundamental to natural science. Theology deals with things that are separable from matter and immovable. The situation with mathematics is less straightforward. Some mathematical studies—arithmetic and geometry—treat things as unchangeable and separable from matter. Other branches of mathematics (e.g. astronomy) deal with things that are unchangeable and material. Because it studies the eternal, astronomy is the mathematical science that is most akin to philosophy. At the same time, some branches of mathematics—optics, harmonics, and also astronomy—share something with natural science. For example, in optics, lines are studied as physical, rather than as mathematical.

Aristotle was keenly aware of the challenges of offering explanations of natural phenomena. Near the beginning of his work on meteorology, he cautions us that with regard to meteorological phenomena, in some cases we are puzzled, and in others we understand things only up to a point. The reader has been warned: the study of meteorology is by no means straightforward, and certain knowledge simply cannot be guaranteed. The caveat regarding the difficulty of the subject appears within a list of phenomena to be investigated (including comets, earthquakes, thunderbolts, and whirlwinds), reinforcing the sense of tentativeness. But whether or not the topics considered are truly inexplicable or 'merely' difficult is an important question. At the beginning of his work *On the Soul*, Aristotle issues a similar warning, that to gain knowledge of the soul is one of the most difficult things in the world. Having said this, he then turns to a discussion about methodology, making it clear that he still expects to be able to do some meaningful work on the topic. In spite of everything, he thought he could make progress in explaining both weather and the soul, subjects that are still challenging today.

The difficulties involved in accounting for some phenomena may be due to problems in accessing information about them. Factors such as distance and difficulty of observation (as in the case of clouds) and rarity of occurrence (such as lunar rainbows) limit our ability to observe certain things. But, once again, just because we can't know everything doesn't mean we can't try to understand: Aristotle is satisfied that we have given a rational enough account of things not fully apparent to the senses if we have produced a theory that accounts for what is possible.

Aristotle's satisfaction with a 'rational enough account' of 'what is possible' may seem surprising in light of some of his discussions of the nature of knowledge, understanding, and explanation. Epistemic knowledge (sometimes translated as 'science') is arranged systematically, with proofs or demonstrations involving deductions from first principles. Demonstration (*apodeixis*) depends on the use of deductive inference, in the form of a syllogism, an argument in which certain things are assumed and something different necessarily follows. For example: if *A* is true for every *B*, and *B* for every *C*, *A* must then be true for every *C*. Aristotle presents his model of scientific knowledge in the *Posterior Analytics*. Scientific explanations require definitions that link what is to be explained (the *explanandum*) to (knowable) first principles, via a deductive framework. Aristotle illustrates the distinction between a definition and a demonstration using these examples: 'What is thunder? Extinction of fire in cloud. Why does it thunder? Because the fire is extinguished in the cloud.' The inclusion of these meteorological examples in the discussion of scientific explanation is somewhat surprising, given his view that accounting for meteorological phenomena is a difficult task.

However—and somewhat notoriously—many of Aristotle's scientific works, including the *Physics*, *On the Heavens*, *Generation of Animals*, *Parts of Animals*, and *Meteorology*, do not obviously conform to the model he lays out in the *Posterior Analytics*: the explanations offered in these are not usually presented syllogistically. Aristotle doesn't seem to have practised what he preached, for the explanations offered in the scientific works do not fit his own prescription of 'how to do explanation'. The relationship between the formal method prescribed by Aristotle and the actual scientific explanations he offered poses questions for those who study his work.

Physics

In the *Physics*, Aristotle identifies four kinds of causes to be used to explain things in nature: the material, formal, efficient, and final. These four causes can be understood as providing answers to four different sorts of questions: 'What is it composed of?', 'What is it?', 'What brought it about?' (or, 'How did it come to be?'), and 'What is it for?' While the identification of these four causes may represent an ideal explanation for Aristotle, the explanations he offers in his own scientific works do not always conform to this model.

According to Aristotle, the world has two distinct regions, the terrestrial and the celestial. Each region is characterized by specific material elements and motions. There are four elements in the terrestrial region—Earth, Water, Air, and Fire—differentiated by the primary contrary qualities of hot, cold, dry, and moist. Natural motions in the terrestrial region are rectilinear, and these four elements either move away from or towards the centre of the cosmos: Fire and Air move away from the centre while Earth and Water move towards it. The physical bodies encountered in everyday life are not the elements, but are each a sort of compound, containing at least two of the elements; it may even be the case that all compound bodies contain all four elements. What we know as everyday, garden-variety 'earth' is a compound composed of the elements Earth, Water, Fire, and Air.

There is a fifth element that is the stuff of the celestial bodies whose natural motion is circular; that element, in the celestial region, is known as *aither* (or *aether*, in Latin). The *aither* varies in purity, being more pure where it is more distant from the terrestrial elements and less pure in closer proximity to them. The four terrestrial elements are themselves material causes, while the motion of the continually moving bodies in the celestial sphere is the efficient cause of some terrestrial events. In other words, there is interaction between the regions, and the celestial region can affect the terrestrial. The motion of the *aither* causes material in the terrestrial realm to be heated. For example, rain is due to the efficient cause of the Sun's motion as it gets closer to or more distant from the Earth, and the subsequent processes of evaporation and condensation.

In the *Physics*, Aristotle asks whether nature acts for the sake of something. He offers an account of rain as water that rises when heated by the Sun, then cools, condenses, and falls to Earth, but is not particularly concerned with explaining rain as such. Rather, he addresses a more general question: how to account for natural objects, events, and processes that appear with regularity. He queries whether the sharpness of front teeth and the bluntness of molars is coincidental and rejects that possibility, because this arrangement occurs with regularity and that regularity is linked to purposefulness. The sharpness of front teeth allows them to tear while the bluntness of those in the back enables them to crush food; both are useful for eating. In Aristotle's view, all natural things happen in a given way and occur for the sake of something; that is the final cause. He emphasizes the teleology of nature, its goal-directedness; this concern with final causes is a signature feature of his natural philosophy.

Living things

While many of his predecessors, including Plato, were interested in the origin of life, prior to Aristotle the study of living things focused primarily on humans. He regarded the study of other animals as part of natural philosophy. He spent time on the island of Lesbos dissecting various animals and gathering information from people with specialist professional knowledge, including fishermen and beekeepers. Aristotle's father, Nicomachus, was court physician to Amyntas III of Macedon and this may have influenced his interest in living things. Aristotle occasionally remarked on areas of overlap between the work of doctors and those who study nature. A number of his works deal with subjects of concern to both physicians and *physiologoi*, such as life and death. Roughly a quarter of what is reckoned to be Aristotle's written work is devoted to the study of animals. And he evidently included the study of animals in his teaching; his writings describe diagrams of animals that could have been used to illustrate lectures. The study of comparative anatomy was part of his overall systematic study of animals, in the *Generation of Animals*, *History of Animals*, and *Parts of Animals* (Figure 9).

Especially in his studies of living things, Aristotle pointed to the reason why things are as they are, the goal for which they have come to be. In *Parts of Animals* Aristotle detailed how to approach the scientific study of animals. There, rather than pointing to the more general methodology described in the *Posterior Analytics*, his advice is specific to the task of zoology.

9. Octopus and other marine creatures depicted on a Roman mosaic from Pompeii, 2nd century BCE.

There are analogical similarities across different kinds of animals: what in the bird is feather and scale in fish. The homogeneous parts of animals that are made of the same composite substance—for example, flesh and bone—are constituted from the primary substances, Earth, Air, Fire, and Water. The heterogeneous parts—such as limbs, faces, and hands—are made of homogeneous parts. The order in which individual parts grow is determined by the final cause, that is, the goal of development.

Aristotle invested a good deal of energy in collecting and sharing information. He advocated specific practices, urging readers to take notes and to systematically collate documents and opinions.

These suggestions reinforce the impression that the Lyceum was focused on enquiry. The collection of the ideas of other, often earlier, thinkers was a feature of the coordinated work that characterized the Lyceum during Aristotle's time. However, Aristotle's emphasis on data collection was not only for scientific work. In antiquity, working with colleagues, he was credited with having gathered together the constitutions of many different city-states, and this collection directly informed his work on politics. Having assembled data—from the constitutions themselves—about the political arrangements adopted in various places, Aristotle and his colleagues used this to inform their own ideas about the best possible constitution for a *polis*. In other words, his 'scientific' method was applicable to other fields of enquiry.

Theophrastus

After his death, the leadership of Aristotle's Lyceum passed to Theophrastus. The school was also known as the Peripatos, named after the walkway around the Lyceum. Members of the school were known as Peripatetics. The method, espoused by Aristotle, of collecting and comparing ideas as part of a larger intellectual project, continued to be a hallmark of the Peripatetic school under Theophrastus. Like Aristotle, his interests were wide-ranging, and he wrote on numerous topics related to the natural world. Diogenes Laertius reports that he produced over 200 works on various subjects. Of these, a relatively small number survive, including one on the principles of nature, another *On the Senses*, and a number of shorter works on specific subjects, including *On Odours* (treating perfumes, including their medicinal properties), *On Fire*, *On Winds*, *On Stones*, *On Fish*, *On Sweat*, *On Fatigue*, and *On Dizziness*; a work on weather signs is also attributed to him. As this list of titles suggests, some of Theophrastus' interests focused on the human body, health, and disease; in fact, both Aristotle and Theophrastus were regarded as medical authorities by later medical writers, including Oribasius in the 4th century CE and Caelius Aurelianus, working probably in the 4th or 5th.

Theophrastus also carved out an important area of enquiry through his study of plants, in two works on botany, the *Enquiry into Plants*, which was apparently used to provide information for the writing of the *Explanations of Plants*. The titles give a sense of his own scientific method: enquiry, followed by explanation. Theophrastus and several other unknown authors produced texts that are essentially lists of the opinions (*doxai*) of natural philosophers and physicians, usually organized by topic, but sometimes by a particular named individual or philosophical school. While the practice of collecting and recording the opinions of earlier thinkers may have begun in the 5th century BCE, during Theophrastus' lifetime this approach became integrated into the pursuit of knowledge about nature, as well as philosophy more broadly.

Opinions count

Aristotle had often set out the opinions of others so that he could argue against them in favour of his own view. He emphasized the use of reputable opinions (*endoxa*) as a starting point for enquiry, as part of his dialectical method, by which we are able to reason from generally accepted opinions on any question. Accordingly, familiarity with accepted opinions is crucial. Modern scholars use the term *placita* ('what it pleases someone to think') literature to refer to a group of ancient texts that preserved opinions for later 'mining' and use. Such texts collect and organize opinions and ideas on a variety of topics, including natural philosophy, mathematics, and medicine. Typically, they do not contain other sorts of information, such as biographical details about the people whose opinions are preserved.

Theophrastus was credited by Diogenes Laertius with producing a work entitled *Physikōn Doxōn*, that is, *Tenets in Natural*

Philosophy, in 16 books. Sadly, it does not survive, and no one is entirely certain what form it took. However, Theophrastus' collection of opinions appears to have been collated in a systematic way, perhaps following a model similar to that used by Aristotle when he organized the collection of constitutions of city-states. The surviving *placita* texts may be excerpts from this lost work of Theophrastus. His collection would have provided a sizeable body of material to be used in philosophical and scientific work by himself, students, colleagues, and later readers, and would have been a very valuable resource. Doxographies—collections of opinions—served their readers as a kind of archive which could function as a resource or tool to aid further work. As today, authors relied on secondary sources, including summaries of the works of earlier thinkers, in handbooks and epitomes. The doxographies are a special type of text produced as part of an intellectual practice that incorporated and built upon earlier work.

Questions and answers

The Peripatetics are associated with another type of text, presented as a series of questions and answers. The work known as the *Problems*, sometimes attributed to Aristotle, is an example of a compilation of such queries and responses, many relating to natural philosophical or medical issues; it may have been produced as a group project. One of a number of texts attributed to Aristotle, it was most likely not written or assembled by him, but rather by members of his school, under the leadership of Theophrastus and his successors. The *Problems* has 38 'books', covering a wide range of subjects, from problems connected with medicine, to others associated with justice and injustice. Nearly 900 questions are organized by topic.

Book 1 deals with problems connected with health and disease, such as: 'Why do great excesses produce disease?' and 'Why do the changes of season and winds intensify or halt, and determine and produce diseases?' Elsewhere in the work, we find problems concerned with mathematics (Book 15) as well as shrubs and plants (Book 20); Book 25 deals with air and Book 26 with wind. The problems are posed in a particular fashion; many begin with the question: 'Why?' The answers are often in the form of a question: 'Is it because...?' The range of topics covered overall is vast, and it is likely that different specific 'books' were copied and made available by subject, so that if an individual reader was only interested in medical topics, for example, they would be able to acquire only the desired portion, and ignore the rest.

These question-and-answer texts may have served as a list of active and interesting problems for enquiry. It is easy to imagine that the questions posed and answers offered reflected known opinions and as such were related to doxographies, also serving a type of archival function to allow the recovery and reuse of earlier ideas. Furthermore, the problem texts offer opportunities to test and apply the principles of Aristotelian science, showing how to explain various phenomena. These two distinctive types of the texts—the collections of opinions and the collections of questions and answers—can be understood as technologies specifically designed to gain some level of control over large amounts of information, enabling easier reproduction of the format used for communication while facilitating the preservation and use of the ideas reported.

Moreover, these Peripatetic texts highlight an approach to philosophizing about nature that emphasized the continuing fascination of certain topics of enquiry and the collaborative nature of a group activity that used methods reliant on data, accumulated and shared over generations. And there may have been another motivation for collecting others' opinions: to produce intellectual histories. Such histories enabled their authors and readers to identify with other thinkers as well as particular philosophical schools, reinforcing the sense of community and membership while also adding prestige through association with intellectual predecessors.

Chapter 7
Old school ties

Membership of and identification with a particular group or school were important for ancient philosophers. For ancient physicians, an association and identification with a medical 'school' or approach also mattered. Even some individuals who might not be regarded, in the first instance, as philosophers made a point of advertising their links to a specific school (including Seneca, who identified as a Stoic). While membership of a group was important, so was personal reputation. In fact, several 'heroic' individuals achieved a sort of celebrity status as mathematicians and philosophers of nature. Two figures who achieved great fame in antiquity were Posidonius of Apamea (in Syria; aka 'Posidonius of Rhodes', where he died, in the middle of the 1st century BCE) and the 3rd-century BCE mathematician Archimedes. The founder of the Athenian philosophical school known as the Garden, Epicurus, who was born in Samos in 341 and died in Athens in 270 BCE, was regarded as a 'hero' or a semi-divine individual by some.

Anecdotal evidence from antiquity describes touristic activities linked to particular 'great men' of science: Cicero reports his own expedition to see the grave of Archimedes. Such activities were part of a culture in which interest in scientific and mathematical work found new modes of expression in the Hellenistic period, including the wider dissemination of scientific ideas and the public display of objects associated with natural philosophy and mathematics.

Ancient Greek philosophers did not always agree about what constituted knowledge or how it could be attained. Some Hellenistic philosophical 'schools' included physics as a part of philosophy, alongside epistemology, ethics, and logic. Questions persisted about what may be known about the world, how knowledge is acquired, and what purpose knowledge serves. Thinkers also drew attention to the limitations on acquiring knowledge. Members of two schools, the Stoics and Epicureans, shared the aim of attaining tranquillity or freedom from worry (*ataraxia*), yet their explanations of the world were often radically different, even opposed. (Modern usages of the term 'Epicurean' tend to focus on pleasure, especially with regard to food, rather than being worry-free.)

Epicurus and the Garden

Diogenes Laertius recounts that Epicurus claimed to have first turned to philosophy at the age of 14, as a reaction prompted by his disgust at his teachers' lack of ability to explain 'Chaos' in Hesiod. He grew up to become the founder of an important school of philosophy, known as the Garden. His school was noteworthy in that he welcomed women. Openness was a special feature of the Garden, and probably contributed to its appeal and the proselytizing zeal of some of its members, not least the Roman poet Lucretius and also Diogenes of Oinoanda, who, in the 2nd century CE erected an enormous stone 'billboard' summarizing Epicurus' teachings, 2.37 metres high and extending about 80 metres. The whole inscription apparently originally contained around 25,000 words.

A chief goal of Epicurus' philosophy was to be free from anxiety. As part of his work to help achieve *ataraxia*, Epicurus developed a materialist philosophy to explain the world, which consists of atoms and empty space (the void). Epicurus denied the possibility that the gods have any influence in the world; he argued that they are too busy being blissful, in the interstices between worlds, to bother with us.

Epicurus was content for his followers to have an 'in principle' approach to understanding the world. He explained that it is not necessary to become too caught up in details. In fact, he argues that less is more, because knowing less can prevent more confusion and worry. In Epicurus' view, there is a distinct possibility of being blinded by science, and this is to be avoided. While Epicurus indicates that he had himself worked out the details of his views more fully, he advocates that a summary is sufficient. A physical theory that can withstand objections and leads to peace of mind should be accepted.

Epicurus wrote a great deal, but only a few short works of his survive. The *Principal Doctrines* is a collection of epistemological and ethical maxims. We also have three letters, meant to be brief summaries of his ideas presented in a convenient—and easily portable—form for his followers. These works survive because they were quoted, in their entirety, in the biography of Epicurus presented by Diogenes Laertius, in his *Lives of Eminent Philosophers*. Epicurus was the final philosopher discussed by Diogenes Laertius in this lengthy work, and he appears to have been the writer's favourite.

Epicurus presents the fundamentals of his physics in the letter addressed to someone named Herodotus (not the earlier historian). Everything (*to pan*) consists of bodies and space; there is nothing else that we can conceive of existing. Knowledge is based on sensory experience and reason relies on sensory information in attempting to infer the unknown from what is known. Bodies are composed of indivisible atoms. Furthermore, the number of bodies is infinite.

In another letter, Epicurus addresses his thoughts to Pythocles, who had asked for a summary of his views. This letter is deliberately brief: it was never meant to replace Epicurus' larger works, for example the (only partially extant) *On Nature*, but instead to serve as an aide-mémoire, emphasizing Epicurus' view that too much detail is not helpful. In the letter, Epicurus explains the *meteora*, the lofty things. Epicurus makes it clear that his primary aim is to provide peace of mind by eliminating fear of meteorological phenomena. To achieve this, he calls for the rejection of traditional ways of explaining weather, including those that affirm divine intervention.

Epicurus rejects the idea that weather events can serve as an indication of any god's action, for this violates his theology. The Epicurean gods are not concerned with human events, and therefore cannot aid with weather prediction. This denial of a causal link between gods and weather signs is emphatic, pointing to what may have been a popular interpretation of such signs in Epicurus' time, as well as the fear evoked by some weather events. Even if divine wrath was not an issue, the effects on agriculture could be devastating. Epicurus argues that even though certain animal behaviours are held as being predictive of weather, these are just coincidence:

> for the animals offer no necessary reason that stormy weather should occur; and no divine being sits observing the coming and goings of these animals and then fulfils their signs. For such folly would not afflict the ordinary being, however little enlightened, let alone one who has attained perfect happiness.

In other words, the gods have better things to do.

Epicurus strongly advocates the use of empirical observation, emphasizing that any explanations offered must not contradict experience. Even though what we see may be explained by a number of causes, none of these explanations may contradict sensory perception. And, even though sense perception contributes to knowledge, Epicurus recognizes that direct perception is not always possible; in such cases, analogy is useful and is one of the principal ways in which we come to understand, for it allows us to extend knowledge from what is encountered via our senses to that which cannot be so perceived. For example, Epicurus suggests that thunder might be due to wind rolling around in the hollows of clouds, in a way that is similar to when air is imprisoned in jars; we cannot access clouds directly, but we are familiar with household objects. The use of analogy is especially important for the explanation of distant phenomena.

Characteristic of Epicurus' approach to the explanation of phenomena is his willingness to entertain the idea that there may be any number of possible causes. He explains that it is impossible to understand all things equally well. Some things—for example, human life and the general principles of physics—are easier to explain than others. For those things that are more difficult, multiple causes and multiple accounts may be offered. Furthermore, Epicurus argues forcefully against being overly attached to any dogmatic explanation. He claims that this is a superstitious trap into which many astronomers and physicists have fallen, warning that 'whenever we admit one argument, yet reject another that is equally consistent with the phenomena, it is clear that we abandon the study of nature entirely and plunge into myth'. Unlike some, he rejects myth as a form of explanation.

The elimination of fear and anxiety—particularly about the intervention of gods in the natural world—motivated Lucretius to present Epicurean philosophy and physical theory in his *On the Nature of Things* (*De rerum natura*), one of the great works of Latin literature. Throughout his six books of hexameter verses, Lucretius aims to show that the basis of the world is material and natural. While human beings sometimes believe that supernatural or divine beings, such as gods, are responsible for the creation and workings of the world, this sort of explanation is unnecessary. Once we understand the material workings of the natural world, there is no need to fear the supernatural intervention of gods.

The Stoa

Epicurus' ideas resonated not only with the members of his school in Athens; through Lucretius' Latin poem, they reached many more people. But the Stoics (named after their school which was known as the 'Stoa', short for the *Stoa poikilē* or Painted Portico/ Colonnade in the Agora of Athens) competed with the Epicureans in the philosophical marketplace. The Stoic worldview, in contrast to that of the Epicureans, emphasized the permeation of the cosmos by the divine. Whereas followers of Epicurus tended to present themselves as being focused on the views of Epicurus himself, the views of members of the Stoa were diverse and not always in agreement.

Our knowledge of the natural philosophy of the early Stoics—who flourished in the 3rd century BCE and included Zeno of Citium (in Cyprus), the founder of the school, and his successors Cleanthes of Assos (in Asia Minor) and Chrysippus of Soli (in Cilicia)—is based on reports preserved by later writers, many of whom were hostile to their views. Only a few fragments of the writings of these Stoics survive, as is also the case for Posidonius, considered by many in antiquity to have been a premier natural philosopher. Unlike the followers of Epicurus, those who aligned themselves with the Stoic school were often engaged in some way in public life, through politics, law, and teaching. Cicero and Seneca are two important Roman authors who tell us a good deal about Stoicism, but Cicero was not himself a Stoic.

Stoic philosophy emphasizes the importance of physics, alongside logic and ethics. Often, our sources attribute ideas to the group ('the Stoics'), rather than to specific, named individuals, and there is sometimes a tendency to homogenize their views. Yet, some sources point to differences in their ideas. Diogenes Laertius lists particular writings by their Stoic authors, so we know of works such as Zeno's *On Substance*, Cleanthes' *On Atoms*, Chrysippus' *Physics*, Archedemus' *On Elements*, and Posidonius' *Discourse on Nature*. The latter author, while regarded as a Stoic, may have been something of a maverick in his thinking.

Posidonius was credited with having written extensively on scientific topics, but only fragments and reports of his writings survive, even though his ideas were widely cited by later ancient authors. He was a student of the 2nd-century BCE Stoic philosopher Panaetius of Rhodes, and wrote on many topics, including logic, ethics, physics, and history. Other ancient authors name titles of about 30 books by him, including *Discourse on Nature*, *On the Cosmos*, *Meteorology*, *On the Size of the Sun*, and *On Gods*. He was recognized for having done work on a range of natural phenomena, in what might (in modern terms) be described as meteorology (which in antiquity included what we regard as seismology), astronomy, mineralogy, hydrology, and geography. Modern scholars have used descriptions of Posidonius' ideas offered by other ancient authors to try to reconstruct his views, but it is sometimes difficult to ascertain which fragments are rightly attributed to him.

A number of concepts are taken to be central to Stoic physics. The cosmos (world) is finite and spherical, a living being ordered by reason and providence; it is surrounded by emptiness, the infinite extra-cosmic void. Time is the measure of the world's motion. According to Stoic physics, there are two principles: 'god' (or 'reason') and 'matter'. 'Matter' is passive, while 'god' is active and exists throughout. These principles (or *archai*) are imperishable and ungenerated, but individual Stoic thinkers identified the rational animating power variously as 'god', a form of Fire, *pneuma*, or breath. The world came to be and will perish, in a fiery conflagration, before coming to be again in a cosmic cycle of regeneration.

One of our sources on Stoic views is Galen, a physician who wrote on philosophical topics. He explains the Stoic conception of *pneuma*, which plays a crucial role in the cosmos. The *pneuma* or breath is made of Fire and Air, and enables the role of the active principle in giving form to matter, which it sustains and in which it exists. *Pneuma* is the underlying cause of the characteristics of things, which differentiate them from one another. According to Plutarch, it is what makes iron hard, stone dense, and silver white. A much later author, Nemesius, the 4th-century CE Bishop of Emesa (in Syria), provides more detail on how this works: the particular characteristics of a thing result from the tension between expansive and contractive forces in the *pneuma*, which has a sort of elasticity.

Cosmological questions

For the Stoics, the natural motion of all matter is towards the centre of the cosmos. While in Epicurean physics, the infinite world is composed of an infinite number of indivisible atoms and infinite empty space in which atoms move, Stoics conceived of the world as being full, a plenum in which matter is continuous and there is no void. Yet, outside this finite world empty space exists. For Stoics, events are predetermined, while determinism is rejected by Epicureans. Lucretius had explained that it is because of the random change possible in an atom's motion—a swerve—that fate is not determined and free will can operate. Questions about determinism and fate were debated during the Hellenistic period and beyond, sometimes linked to an increased interest in astrology.

The cosmological theory of each philosophical school made reference to theology. Epicurean gods reside in spaces between the infinite number of worlds (cosmoi), uninterested in human affairs. There is no need to worry about them interfering with our world. The existence of various explanations of the same phenomenon should persuade us that phenomenon can be explained naturally, without positing divine intervention. In contrast, for the Stoics, there is only one world, in which matter is passive and acted upon by a divine (eternal) principle. This divinity permeates the entire cosmos, which is itself alive. Epicureans, too, thought of the cosmos—in fact, all cosmoi—as living, and having a 'life cycle'. In spite of their profound differences, Epicurean and Stoic natural philosophy shared some similar ideas, and both schools were especially influential in the early modern period (*c*.1400–*c*.1800), as evidenced, for example, by renewed interest in atomist theories.

The ideas of Epicurean and Stoic philosophy were communicated in poems, as well as in prose texts. Twice in his poem, Lucretius said that he thought using poetry as a medium of communication might make his subject—Epicurean philosophy—more palatable to his readers. He drew an analogy to the way in which physicians coat the rims of cups with honey, to persuade children to take their medicine. The poems of Lucretius and Manilius conveyed, respectively, Epicurean and Stoic ideas about nature and science to ancient audiences as well as later readers and translators.

Chapter 8
Roman nature

It is something of a commonplace to claim that the Romans were better at technology (controlling nature) than science (explaining nature). However, a striking number of Latin authors devoted themselves to writing scientific texts that attracted a wide readership, even well into the modern period. The ideas of many Greek philosophers were taken up, discussed, developed, and spread by Roman authors including Lucretius, Cicero, Seneca, and Pliny the Elder, to name but a few. Some bilingual Roman authors translated Greek works (including Aratus' astronomical poem) into Latin, for example Cicero, whose translation is largely lost.

Some of these Latin works were intended for non-specialist audiences, effectively serving to 'popularize' science. While many educated Romans read Greek, Cicero is a good example of an author who sought to make Greek philosophy accessible to Latin readers through various of his writings, including dialogues that featured discussions of natural philosophical topics. Latin translations of Greek works, like Aratus', indicate that there was a market for natural philosophical and astronomical texts in Latin, among other scientific topics. Lucretius' poem *De rerum natura* introduced Epicurus' natural philosophy to a Roman audience.

A number of authors continued an earlier practice of writing what may be regarded as histories of science, including Seneca in his

Natural Questions and Vitruvius, whose work *On Architecture* also contains accounts of the history of explaining nature, and the invention of sundials. In his *Natural History*, Pliny the Elder depicts a history of striving to understand natural phenomena.

Seneca the Younger

Many Roman works on nature have a strong moral tone, in some instances criticizing aspects of Greek culture while in others showing admiration. Seneca's *Natural Questions* discussed meteorological theories, revealing good knowledge of Greek ideas and also conveying ethical messages. Seneca was the prolific author of many works still extant, including tragedies, satire, and philosophical treatises. During his lifetime he was politically active and, at times, influential; for example, he was tutor to the young Nero and later one of the Emperor's advisers. His writings cover a wide range of topics, but his primary aim was to inspire ethical improvement.

Seneca retired from public life in 62 CE and focused on writing and philosophy. It was during this period, towards the end of his life (prior to his suicide forced by Nero in 65 CE, for alleged participation in a conspiracy), that Seneca wrote the *Natural Questions*, in which he provided a detailed discussion of meteorological phenomena. The work does not survive in its entirety, and it is difficult to determine the original order of the books and their topics. Several of the extant books contain prefaces and epilogues in which Seneca provides moral exhortation; the *Natural Questions* fits with his overall programme of improvement. While this was an important motivation, Seneca also gives evidence of his long-standing interest in natural phenomena. He mentions in the *Natural Questions* that as a young man he wrote a work, now lost, on earthquakes. Even as his meteorological writings may have been part of a larger programme designed to aid in ethical improvement, Seneca's fascination with meteorology as a subject is clear.

Seneca's approach to explaining natural phenomena shares much with that of his predecessors, whose ideas influenced his choice of topics. By and large, he addressed similar subjects to those covered by Aristotle in his *Meteorology*. As Aristotle, Theophrastus, Epicurus, and Lucretius had done before him, in his explanations of natural phenomena Seneca employed analogies to familiar experiences, some of which in his case gave details about contemporary Roman life (including the fashion for using snow to chill drinks). Seneca was particularly interested in etymology and aimed to present clear definitions of relevant terminology. He provided historical information on the natural explanation of phenomena, as he discussed and criticized others' explanations in the course of arguing for his own views. He shared information on many thinkers, including Posidonius, whose works, as mentioned previously, are largely lost.

For Seneca, the study of the physical world is not separable from human concerns and activities, including morality. He encouraged others to study nature in order to leave behind sordid things, to keep the mind separate from the body, and to exercise the mind on obscure matters, so as to successfully deal with ordinary ones.

The *Natural Questions* is addressed to Seneca's close friend Lucilius, who, in turn, has sometimes been identified as a possible author of *Aetna*, a poem about volcanic activity. Given the volcanic character of parts of the Mediterranean world, it is perhaps surprising that *Aetna* is one of few ancient works dealing with volcanoes. Whoever he was, the *Aetna* poet urges devotion to one's parents, while arguing that volcanoes occur naturally and are not due to the influence of the gods.

Pliny the Elder

The Roman agricultural writers espoused the Roman ideal of the 'man of the earth', with feet planted firmly on the ground; they included Cato the Elder, who was active in the political and cultural life of Rome in the first half of the 2nd century BCE and credited by Columella for having 'first taught agriculture to speak Latin'. Varro, writing in the 1st century BCE, and Columella, in the 1st century CE, each presented agricultural calendars, the heritage of which may go back to Hesiod's *Works and Days*. The presumption was that by being familiar with astronomical events, farmers would know which times in the year would be particularly appropriate and favourable for various tasks. Pliny the Elder dedicates a central book of the *Natural History* to the subject (Book 18), self-consciously looking back to Hesiod, whom he regarded as the 'father' of agriculture.

Nevertheless, the Roman relationship with Greek learning and achievement was ambivalent. Pliny, who revered Hesiod, also counselled Roman farmers that they should not rely only on authoritative astronomers but must depend on their own eyes and observations. Book 18 of his *Natural History* contains a sort of farmer's almanac, with instructions on how to construct a wind-rose, to determine wind direction. Pliny made it clear that while astronomical knowledge can aid farmers in predicting the weather, they did not need to be blindly reliant on astronomical expertise. Nevertheless, he described stellar risings and settings, demonstrating that he did not wish to disassociate himself entirely from what is valuable in Greek-derived astronomy. However, he was clearly unwilling to rely solely on the advice of 'expert' astronomers, and emphasized the practical necessity of careful observations being made in individual and specific locations.

The Roman poet Virgil also emphasizes the importance of becoming familiar with the seasons and the weather, and of using astronomical knowledge. He explains

> What's more, you need to keep a weather eye on sky
> formations—
> such as Arcturus, the twin kids of the Charioteer, or Draco, that
> bright light,

and stay vigilant as those mariners who, homeward bound, ride
    stormy seas,
yet venture close to Pontus, the Straits of Abydos and their
    oyster beds.
And when September's equinox doles to day as many hours as
    to night
and splits the world in two fair halves, both equal light and
    dark,
then set to work the oxen, men, broadcast barley in the fields,
until midwinter's whelming showers slap you in the face.

Virgil served, to some extent, as another role model for Pliny, who
cited the Roman poet as an authority in his presentation of his
own farming calendar. Pliny notes, referring to the passage just
quoted, that Virgil gives the advice that 'before any thing else, we
should learn the theory of the winds, and the revolutions of the
stars; for, as he says, the agriculturist, no less than the mariner,
should regulate his movements thereby'. Pliny emphasizes that

it is an arduous attempt, and almost beyond all hope of success, to
make an endeavour to introduce the divine science of the heavens to
the uninformed mind of the rustic; still, however, with a view to
such vast practical results as must be derived from this kind of
knowledge, I shall make the attempt.

This sentiment conveys his sense of duty, to serve his fellow
Romans by sharing knowledge.

Having consulted some 2,000 writings, including Greek, Roman,
and other material, Pliny produced an encyclopedic work covering
a wide range of scientific topics, including astronomy. He
demonstrated that one way of controlling nature, or *natura*, was
through organizing knowledge about it. The ability to manage
knowledge of nature so that it is useful to humans was, in Pliny's
eyes, a singularly Roman achievement, emblematic of the
Empire itself.

While Pliny was keen to celebrate Roman achievement, he complained about the lack of willingness among his contemporaries to undertake original work on understanding winds, despairing that in his time, 'in the blessed peace which we enjoy, under a prince who so greatly encourages the advancement of the arts, no new inquiries are set on foot, nor do we even make ourselves thoroughly masters of the discoveries of the ancients'. He contrasts the scholars of earlier times and other cultures who pursued the study of winds for intellectual reasons with people of his own day interested only in increasing their wealth.

Written in the stars

Familiarity with Greek science was displayed by some Romans hoping to impress. The architect Vitruvius included brief accounts of time-finding devices, as well as natural philosophy and astrology. Stone sundials appear to have been widespread in the Graeco-Roman world: about 500 examples survive. Historians have wondered whether some ancient sundials were merely decorative, adorning the gardens, for example, of Pompeii, as a number of them do not seem to have been constructed for accurate time-finding. Yet, even in antiquity these objects were evidence of scientific and mathematical expertise: Vitruvius offered a brief history of the invention of distinct sundial designs.

While many of these 'garden' dials would have adorned private villas, there are also examples of large time-finding instruments on public display, notably the meridian line or sundial (the so-called Solarium, or Horologium, Augusti) on the Campus Martius in Rome and, as has been argued by modern scholars, the Pantheon itself. Such public displays involving the application of mathematical techniques to track a natural phenomenon (sunlight, indicating time of day and solar year) may have contributed to or even been the result of a wider interest in science and mathematics.

Greek and Roman calendars used apparent cycles of the Sun, Moon, and stars to indicate the times for religious festivals and agricultural activities. Lunar months (of 29 or 30 days, typically from one new moon to the next) were widespread in ancient societies, offering a useful way of marking time. However, a lunar year of 12 lunar months (354 days on average) does not fit well with a solar year (of about 365) aligned with the seasons. Over time, festivals meant to occur in a particular month would occur at the wrong time of the year, according to the solar calendar. By the 50s BCE, the Roman Republican calendar (of 355 days with an intercalary month of 22 or 23 days) was 90 days out from the seasons. Julius Caesar sponsored calendar reform, promulgated through the eponymous Julian calendar. Pliny reports that the expert Sosigenes assisted Caesar. The traditional Roman quasi-lunar calendar was abandoned in favour of a solar one with 365 days and an intercalary day every four years (our modern 'leap' year). While we may not think of calendar reform as a scientific activity, it was important for weather prediction and astrology, among other applications. Sosigenes may also have worked on astrometeorological calendars (known as *parapegmata*). In the early modern period, calendar reform was also undertaken by specialist astronomers.

That interest in astronomy was not confined only to specialists is clear from Manilius' lengthy Latin poem *Astronomica*. Manilius thought that the divine spirit pervades the entire cosmos; the cosmos reflects the *ratio* (reasoning/rationale) of the divine will. In fact, the unity of the natural world and its dependence on the divine (eternal) spirit is his underlying theme, reflecting Stoic ideas regarding the divinity of the cosmos. Manilius was also clearly familiar with Lucretius' *On the Nature of Things*. Indeed, Manilius' poem may have been intended, at least in part, as an attack on Lucretius' Epicurean ideas. Manilius' insistence on the role of the divine spirit actively at work throughout the cosmos contrasts strongly with the Epicurean view. For Manilius, the cosmos is a living thing, requiring nourishment provided by the divine. He emphasized the unity of nature, the links between the celestial, the terrestrial, and even the subterranean. In their turn, the rains, winds, seas, and rivers participate in the process of nourishment, feeding other parts of the living cosmos.

Several features of Stoic philosophy, including those espoused by Manilius, fit well with the assumptions of astrology. The interconnectedness of the entire cosmos is vital, in several senses, not least because the cosmos is a living organism. What happens in one part of the cosmos has influence elsewhere.

Astrology had political as well as personal import in Roman culture. The emergence of astrology in Rome as a useful tool for political messaging may be linked to the collapse of the Republic and the rise of the Empire. Indeed, imperial propaganda used astrology in various ways, making reference to the zodiac as well as horoscopes and even to the cosmos itself. Perhaps the most important imperial astrological project involved Augustus Caesar, the first Roman emperor, whose public persona was enhanced by the significance of his astrological sign, promulgating the idea of the rebirth of Rome: Capricorn. Augustus had coins struck with the Capricorn to promote his image (Figure 10); several ancient authors attest that he published his horoscope, making official his astrologically predicted destiny.

Roman astrology incorporated elements of science and natural philosophy, and also mythology. Greek influences are seen in the Latin works that survive—Manilius' astronomical-astrological poem and the manual of astrology produced in the 4th century CE by Firmicus Maternus. Firmicus' manual attests to the practice of the working astrologer, who may have used various tools of the trade, such as zodiacal 'boards', perhaps to show planetary alignments and to illustrate personal natal (birth) horoscopes.

10. **Capricorn featured on coin of Augustus, 27 BCE.**

An astrologer's toolkit

The so-called Tabula Bianchini (named after the astronomer, historian, and archaeologist Francesco Bianchini, 1662–1729) in the collection of the Louvre dates to the 2nd century CE (Figure 11). Understood to be a zodiacal board, it was clearly a luxury item. As it is made of marble, it would probably have been in a fixed location, requiring clients to come to the astrologer to consult about horoscopes. Inscribed gems, with signs of the zodiac and planets (including the Sun and Moon), also survive from the period. These may have been used with such horoscopic display boards.

Displays of nature

In addition to horoscopic boards and massive marble sundials, a number of striking pieces of art attest to Roman interest in the natural world. The so-called Farnese Atlas, a marble sculpture of the god Atlas holding a celestial globe illustrating constellations, is a Roman copy of an earlier Greek example. It is housed in the National Archaeological Museum of Naples (Museo Archeologico

11. The Tabula Bianchini, 2nd century CE.

Nazionale di Napoli), whose collection also includes numerous frescos and mosaics that depict natural phenomena—including different kinds of animals. The Farnese Atlas does not show individual stars and the celestial circles are inexact, suggesting that the globe itself was meant to be evocative of astronomical and cosmological ideas, rather than a scientific tool.

Decorative mosaics depict familiar animals, such as dogs, and also others less well known. Some animals, such as those on the farm, were discussed by agricultural writers, including Columella, who devoted separate books of his *On Agriculture* to the husbandry of large and small animals. Unusual creatures were objects of curiosity and an attraction. Exotic animals, such as the rhinoceros,

12. Rhinoceros on Roman imperial coin, minted sometime between 84 and 90 CE, during the reign of Domitian.

were depicted on coins, and displayed in public and private collections (Figure 12). The first giraffe to have been brought to Rome was apparently included in Julius Caesar's triumphal public games, in 46 BCE. The capture and exhibition of wild animals afforded Romans yet another way to demonstrate their command of nature.

Chapter 9
River deep, mountain high

In 1927, the distinguished historian of medieval science and magic Lynn Thorndike asked other historians of science: 'When, where, and how did exact measurement of mountains, by scientific method and on a cooperative basis, begin?' He emphasized the 'scientific' and collaborative approach, hallmarks of modern science and also ancient, in some cases. An intriguing question, in response to which the distinguished historian of mathematics Florian Cajori (1859–1930) wrote a scholarly article, addressing the history of the determination of heights of mountains. One of Aristotle's students, Dicaearchus of Messana, working in the early 4th century BCE, may have written the first work on the height of mountains; his work was reported by later ancient authors, including Pliny the Elder.

Estimates of mountain heights, sea depths, the distances to the Sun and Moon, and the sizes of the Earth and the cosmos were offered by many Greek and Latin authors. Aristotle commented on calculations made by others of the sizes of astronomical bodies, suggesting that they are much larger than they appear; he gave a figure of 400,000 stades for the circumference of the Earth. In the 3rd century BCE, Aristarchus of Samos produced a work on the distances to the Sun and Moon, as well as their sizes. In the following century, Hipparchus tried to determine the distance to the Moon by estimating that of the Sun; the 2nd-century CE astronomer Claudius Ptolemy criticized the method he used. Posidonius is also reported to have given figures for the distance (by Pliny, who complains that he doesn't quite follow) and size of the Sun, as well as a measurement of the circumference of the Earth. Eratosthenes, who worked in the 3rd century BCE, is well known for his estimation of the circumference of the Earth (252,000 stades, according to Pliny, or 250,000, according to Cleomedes, writing around 200 CE), and his geometrical deductive method for finding it, based on assumptions that shadows cast in all places are parallel and that the city of Alexandria (in present-day Egypt) is 5,000 stades north of Syene, measured along a meridian. (Discussions of 'sizes' often included the volumes of heavenly bodies; according to Ptolemy, the volume of the Sun is about 170 times that of the Earth. Such calculations highlight how different ancient ideas about these quantifications are from our own.)

Many of those who wrote on such topics had specialist mathematical skills and knowledge, but questions about the sizes and distances of features of the world fascinated others too. Pliny the Elder provides an account of one attempt to estimate the Earth's circumference. He reports that on the death of Dionysodorus of Melos, a celebrated geometrician, his female relatives found a letter in his tomb signed by him and addressed to those on Earth, explaining that he had passed from his tomb to the bottom of the Earth, and found the distance to be 42,000 stades. According to Pliny, other mathematicians took this to mean that the letter had been sent from the centre of the Earth itself, and that the distance represented the longest space downwards from the surface of the Earth to its centre. From this assumption, the mathematicians calculated the circumference of the Earth as 252,000 stades. This calculation, based on a fictional journey to the centre of the Earth, was strikingly similar to that obtained by other means, including that of Eratosthenes. Furthermore, Pliny's description of an imagined scientific exploration shares features with other writings, such as the

Taking the measure

A 'stade', as a measure of distance, represents the length of a running track. In ancient Greece, foot-race courses in different places were not always precisely the same length; furthermore, estimates of the lengths vary. At different places and times, a track (*stadion*) might be longer or shorter. By modern standards, the *stadion* at Delphi may have been about 177.5 metres, the one at Olympia 192 (though other estimates are given), and Miletus 192. The exact length of a stade is often unknown. For this reason, figures given for a stade may vary from about 150 to 210 metres. Herodotus tells us that a *stadion* is six *plethra*; a *plethron* was equivalent to 100 feet. However, Greek 'feet' (foot = *pous*; plural = *podes*) were not all the same length; like other measures, the length of a foot varied by place and time. Some modern accounts of the stade simply offer 200 metres as an approximation.

We do not know the length of the stade Eratosthenes used for his calculation of the circumference of the Earth, and a number of possibilities have been proposed. The American Physical Society notes that values of between 500 and about 600 feet for a stade would result in a calculated circumference of between about 24,000 and 29,000 miles. Compare this to the modern measurement of the Earth: about 24,900 miles around the equator, slightly less around the poles. By another calculation, the Attic stade (from the region of Attica, including Athens; this may have been used by Eratosthenes) of 184.98 metres × 252,000 (the number of stades mentioned by Pliny) gives roughly 46,615 kilometres, compared to NASA's figure of 40,070.

humorous report of a lunar expedition written by Lucian of Samosata, a 2nd-century CE Greek author originally from the Roman province of Syria.

Pliny and Lucian were writing for a broad, non-specialist readership. It is only through their writings that we have access to some of the scientific ideas of the time that would otherwise have been lost. Their discussion of these topics indicates that they were of interest to a wider audience. Pliny, in particular, is an important source for our knowledge of the work of many whose writings do not survive. He reports Posidonius' views about the height at which winds and clouds occur, the height of mountains, and the depths of the sea, the sorts of things we can imagine all kinds of people—not just distinguished philosophers and mathematicians—contemplating, if not actually attempting to measure and calculate.

Mountains

Mountains were of consequence for many reasons, and there were various efforts to estimate or even measure their height. All of the named mountains for which heights are given by ancient authors are in Greece—Cyllene, Pelion, Olympus, Atabyris (on Rhodes), and Acrocorinth. Several authors report Dicaearchus' measurements. Pliny the Elder, in the 1st century CE, explained that he had achieved his measurement of Mount Pelion with royal support; the figure he gave for its height was 1,250 paces. Earlier, the 1st-century BCE author Geminus, in his *Introduction to the Phenomena*, offered Dicaearchus' figures to bolster his argument that clouds and winds form at a height lower than the summits of mountains. Writing in the 1st century BCE, the Greek author Strabo mentions the height of Cyllene in the Peloponnesus, reporting that some give its height as 20 stades, others 15.

Instruments appear to have been used in the determination of mountain heights. Plutarch, in his *Life of Aemilius Paulus*, describes how Aemilius halted below the temple to Apollo Pythius, the Pythium, to allow his army to rest before battle. Plutarch reports that from there, Olympus rises to a height of greater than 10 stades, as proclaimed in an inscription by Xenagoras who measured it. He provides what is apparently a quotation of that inscription:

> As found by perpendicular measure, the height of the peak
> Of the sacred mountain Olympus, at Apollo's Pythium,
> Is a full ten stades, plus four feet short of a plethron.
> Xenagoras, Eumelus' son, measured the distance.
> Greetings, Lord Apollo! May you be generous and merciful!

Plutarch noted that whereas mathematicians claim that no mountain has a height and no sea a depth of greater than 10 stades, it appears that Xenagoras took his measurement carefully, with instruments.

Plutarch doesn't specify which instruments were used. We know from Heron of Alexandria's work *Dioptra* how an instrument of the same name—a sighting tube or rod with sights—could be used with geometrical methods to determine a mountain's height from some distance away. Plutarch's report provides evidence not only about the measurement of mountains but also about the sort of scientific and technical achievements that were celebrated with monuments and inscriptions. And we see from Plutarch's account that the opinions of 'experts'—in this case, mathematicians—were sometimes refuted.

Bodies of water

In addition to the heights of mountains, the depths of rivers and seas were also estimated and, apparently, measured. Strabo reports that, according to Posidonius, the Sardinian Sea is the deepest of those seas that have been measured, about 1,000 fathoms. Other authors offer similar figures for sea depths:

Plutarch mentions 10 stades as a figure given for the deepest, while Papirius Fabianus—as reported by Pliny, who regarded him as very knowledgeable about natural phenomena—prefers 15 stades. Strabo implies that measurements of deep-sea depths were actually undertaken, but it is not known how these figures—which appear to have been rounded—were achieved. (Strabo uses an *orguia* (translated here as 'fathom') as a measure, equivalent to the length of outstretched arms, or about six feet.)

The distinction between what might be regarded as purely theoretical questions and more mundane matters was not always clear. Some, including Archimedes, were interested in both. Estimates and measurements of mountain heights and sea depths were often motivated by practical, including military, concerns. Measurements and estimates of geographical features also informed physics and cosmology, offering answers to questions about the natural world, such as the height at which clouds occur (relevant to where the gods reside, on a cloudless mountaintop, according to Homer) and the sphericity of the Earth.

The Earth and the cosmos

While many educated Greeks and Romans would have thought of the Earth as spherical, mountains rising above the plains made clear that the Earth was not perfectly round. Estimations and calculations of mountain heights gave credence to the argument that their altitude did not alter the essential sphericity of the Earth. Cleomedes maintained that there are no mountains with a height of greater than 15 stades, nor a sea of such depth. He explained that because the diameter of the Earth is greater than 80,000 stades (according to Eratosthenes, as he noted), a protrusion of even 30 stades would only appear to be as a speck of dust on a ball. Such relatively minor deviations on its surface do not contradict the Earth's sphericity. Some have suggested that Epicurus and his followers thought that the Earth is flat, but there is no evidence to support this view and, furthermore, we would not expect the Epicureans necessarily to have had a firm view about the shape of the Earth.

Ideas about the shape and size of the cosmos impacted views about the order of the cosmos itself. Aristarchus of Samos was reported, by Archimedes in *The Sand-Reckoner*, to have considered the Sun to be in the centre. His vocabulary (radius, circumference, etc.) attests to his preoccupation with mathematics:

> Now you are aware that the 'universe' is the name given by most astronomers to the sphere the centre of which is the centre of the earth and whose radius is equal to the straight line between the centre of the sun and the centre of the earth. This is the common account given by astronomers. But Aristarchus of Samos brought out a book consisting of some hypotheses, in which the premisses lead to the result that the universe is many times greater than that now so called. His hypotheses are that the fixed stars and the sun remain unmoved, that the earth revolves about the sun in the circumference of a circle.

Using geometry to describe the world order, Aristarchus entertained the possibility of the Earth being in motion around the Sun. Vitruvius credits him as the inventor of two different types of sundials—including the *skaphe* (bowl) dial—tracking the apparent motion of the Sun, to tell the time of day and year (Figure 13). Other mathematicians were also named as the inventors of specific kinds of sundials.

Mathematicians and philosophers not only counted and measured the features of the natural world: they also produced tangible, material models of various phenomena. Cicero regarded Archimedes, a mathematician of the greatest renown in antiquity, as the founder of the branch of mechanics devoted to the construction of planetaria, depicting the motions of the wandering stars. Several other ancient authors also mention Archimedes' models. According to Pappus of Alexandria,

13. Sundial from Delos, Greece; date unknown.

writing in the 4th century CE, Archimedes wrote a work *On Sphere-making*; it does not survive. Posidonius was also credited with having built a planetary model. In the 2nd century CE, both the astronomer Ptolemy and the physician Galen referred to models of planetary motion and their makers. None of these models are extant but the accounts of ancient planetaria served to intrigue and inspire others, well into the early modern period. Both Petrarch (Francesco Petrarca), in the 14th century, and Marsilio Ficino, in the 15th, made approving references to Archimedes as a planetarium-maker, having produced a physical model of the cosmos. The so-called Antikythera Mechanism was a portable astronomical teaching device, possibly dating from the 1st century BCE; all that remains of it are 80-odd corroded bronze fragments, some of which have gears while others are engraved in Greek.

Big numbers

We find a special fascination with very large numbers in writings on a range of topics. Aristarchus and Eratosthenes contemplated

The Antikythera Mechanism

In 1901, sponge divers working near the Greek island of Antikythera found a mass of bronze from a shipwreck which may have occurred around 60 BCE. The divers' find has fascinated archaeologists, historians of science, mathematicians, and astronomers. Originally it displayed various calendars and astronomical information, allowing it to be used in teaching about astronomical motions and predicting eclipses, while also providing the dates of the Panhellenic games and the equinoxes. The device is the oldest known mechanical calculator and can be understood as a 'masterpiece': it was very probably one of a kind. Sophisticated modern forensic techniques—including radiography and surface imaging—and the knowledge of historians and mathematicians have enabled detailed understanding of the device.

the size of the cosmos and the number of prime numbers. Archimedes calculated the number of cattle belonging to the Sun (in the *Cattle Problem*, addressed to Eratosthenes, recalling Helios' herds in the *Odyssey*) and the number of grains of sand required to fill the cosmos.

Pliny the Elder proudly provided the details of the very large number of sources he consulted while compiling his *Natural History*. The first book of his work is actually a very dense table of contents, listing the subjects covered in the 36 books that follow. For each book, Pliny outlines the topics discussed as well as the sources he consulted; he also reports the total number of facts, descriptions, and observations offered in each. Book 2, which deals with cosmology and astronomy as well as other subjects, conveys 417 facts, observations, and investigations. Book 6 is concerned with various places and lists 1,195 towns, 576 clans (or races; *gentes*), 115 famous streams or rivers, 38 famous mountains, 108 islands, 95 obsolete or extinct towns and races, as well as 2,214 facts, enquiries, and observations. Pliny celebrated minutely detailed—and quantified—knowledge of the world. Note that he does not use round numbers. The implied precision of his counting conveys a further sense of detail and authority. One has the impression that—for Pliny and possibly also his intended readers—more is actually more.

Many of the works dealing with large numbers are lost, but four survive, by two authors both recognized as specialist mathematicians: Aristarchus' *On the Sizes and Distances of the Sun and Moon* and Archimedes' *Measurement of the Circle, The Sand-Reckoner*, and the *Cattle Problem*. The latter two works, in particular, appear to have been written with non-specialist readers in mind. *The Sand-Reckoner*—concerned with how to express very large numbers—was addressed to the Syracusan king Gelon, Archimedes' presumed patron, but not himself a mathematician. The *Cattle Problem* was presented as an epigrammatic poem, noteworthy for the difficulty of the mathematics involved, in which a seemingly simple and practical problem about counting herds of different coloured cattle is set out in metre. The problem is posed almost as a riddle, and no solution is offered.

Scholars have questioned whether Archimedes was the author of the poem (in fact, the problem-poem may have been written and presented by someone else), but most assume that he was familiar with the problem presented there. The problem requires eight unknown quantities to be found. It was satisfactorily solved only in the 20th century, and the solution required the use of computers, resulting in a very large number: more than 100,000 digits. We don't know whether the ancient author of the poem thought that it was solvable and, if so, by what means. The text we have opens this way: 'a Problem (*problēma*) which Archimedes devised in epigrams (*epigrammata*), and which he communicated to students of such matters at Alexandria in a letter to

Eratosthenes of Cyrene'. Epigrams were regarded as a special type of poetry; typically concise and brief, they were used to mark something (including an event) that was particularly noteworthy.

A number of letters written by ancient Greek mathematicians survive. Some of these are clearly communications between friends and colleagues, and have almost the flavour of a conversation, while others offer a challenge to be solved. Archimedes wrote to various individuals interested in mathematics; he apparently often sent out enunciations without proofs, that is, puzzles in advance of the works themselves. We know from other sources that Archimedes, living in Syracuse, corresponded with Eratosthenes of Cyrene, the Librarian in Alexandria. (The Museum and the Library at Alexandria were established under Hellenistic rulers.)

Eratosthenes was a polymath, a scholar active in different areas, with the nickname 'Beta' ('second', because in every branch of knowledge he was brilliant, but not number one). Even if not 'first', he made original contributions on a range of topics, including understanding of the Homeric poems. He is credited with having devised a mechanical solution to the problem of doubling the cube, constructing a cube whose volume is twice that of a given cube. This was one of the three 'classical' problems of ancient Greek mathematics, along with squaring the circle (constructing a square equal in area to that of a given circle) and trisecting an angle. Eratosthenes presented his solution in an epigram, conveyed in a letter to his patron, King Ptolemy (possibly Ptolemy III), further evidence that mathematical problems and solutions were sometimes shared via poetry. Nevertheless, we are left to wonder why mathematical problems were communicated via poems. Might Archimedes and Eratosthenes—if they actually composed the poems attributed to them—have been showing off their range of talents and interests, rather like (later) Renaissance men?

Chapter 10
Is there scientific progress?

In the 2nd century CE, the mathematician Claudius Ptolemy and the physician Galen of Pergamum made contributions to scientific and medical thought and practice that remained foundational for well over 1,000 years. Both regarded themselves as innovators and educators, building on the work of their predecessors and synthesizing information for the benefit of others, including students and practitioners wishing for more knowledge. They discussed the proper methods for astronomy and medicine respectively and highlighted their own ambitions to produce new work that would have lasting value. Not only scientific and medical experts, they were also philosophers of science and medicine, probing the basis of knowledge in their fields, concerned with the foundations, methods, and implications of their work.

Ptolemy and Galen both refer to the work of predecessors, making it clear that it provided a basis for their own endeavours. At one point, Ptolemy explains that he will recount what has been established adequately by 'the ancients', pointing to areas in which there was disagreement, for example in the length of the year. He takes into account the observational work of the ancient astronomers, admiring in particular Hipparchus, whom he describes as having been industrious and a lover of truth, and whose own work—particularly on the precession of the equinoxes—relied on astronomical data accumulated over a long period of time. For some of his observational data, Ptolemy was heir to a number of traditions, including Babylonian astronomy as well as Greek; his astronomical calendar used Egyptian month names.

Galen, who was a prolific author and famous in his own time, devoted a lengthy work to arguing that his two great intellectual heroes—Hippocrates of Cos and Plato—were largely in agreement in their ideas on nature, physics, and the soul. He used their opinions as a basis for his own arguments, a tactic similar to the use of opinion in Aristotelian dialectic. Galen also wished to show that he understood the views of Hippocrates and Plato better than others did.

Ptolemy wrote on a wide range of topics, including geography, harmonics, astronomy, astrology, optics, and scientific instruments. While his ideas and methods have been studied for centuries, in the past century most attention has been paid to the

Hippocrates of Cos and the 'Hippocratic writings'

Hippocrates of Cos, who was probably a contemporary of Socrates, was one of many doctors and medical writers working and sharing ideas in the late 5th and 4th centuries BCE. Some of their work survives, but not always in their entirety. We know that a physician named Hippocrates existed, but we know very little about him, or which and how many of the approximately 60 works attributed to him in antiquity were actually composed by him. What survives with an attribution to Hippocrates varies greatly in style, background, and doctrine, and we don't know how many different authors were actually responsible. The term 'Hippocratic writings' is used to refer to works attributed to Hippocrates in the ancient or early medieval periods.

mathematical aspect of his writings. In fact, Ptolemy himself emphasized the importance of mathematics, as he was convinced that mathematics offered the best means to understand the world. But he did not confine himself to theoretical ideas. He had a hands-on approach to scientific and mathematical work, describing in detail several scientific instruments, including those of his own design; some were for astronomical observation, others for harmonic experimentation. He even included instructions for making and using instruments, including an astronomical globe. He specifically mentioned Alexandria as a site from which he made observations; it was also the location he used to establish the times of the positions of the astronomical bodies.

Galen was also engaged in a hands-on manner, working as a physician. He repeatedly stressed the importance of empirical information for physicians, especially emphasizing the crucial nature of anatomical knowledge. In his own case, much of this was acquired though his experience of dissection, including that of an elephant's heart procured from Caesar's cooks. He also conducted experiments on animals, including pigs. For both Ptolemy and Galen, their work depended on using theoretical and practical knowledge.

Ptolemy's mathematics

Indeed, for Ptolemy, mathematics qualified both as practical and theoretical philosophy. And, while he approvingly cites Aristotle at the beginning of his great astronomical work, he turns Aristotle's hierarchy of knowledge on its head. For Ptolemy, mathematics was the highest form of philosophy, and mathematicians the greatest philosophers. He argued that by studying, teaching, and advancing theoretical astronomy, a person becomes as divine as is humanly possible. His emphasis on the value of mathematics for understanding the world gained a devoted following in scientific circles, even in later periods—particularly among mathematicians. Ptolemy was convinced that there are profound interconnections throughout the cosmos; this helps us to understand his commitment to studying different aspects of the world.

Ptolemy's *Harmonics* was as influential as his astronomical works, and was linked to his ideas about the cosmos. He conducted experiments on sound using instruments of his own invention, and shared the design and his results. Describing sounds mathematically, he drew analogies between astronomical bodies, harmonic structures, and human character. For Ptolemy, the beauty of astronomy and harmonics—both mathematical pursuits—was a direct reflection of the beauty and order of the cosmos, which the study of astronomy transmits to our souls. Ethical improvement is possible because the structure of the cosmos enables influences (and analogies) between the celestial and the earthly, between the divine and the human. By studying these relations—embodied in the physical world and perceived visually in the sky above and aurally through musical sounds—and by describing them mathematically, in principle everyone can achieve some measure of the divinity present in the cosmos, improving themselves and their lives. We can become better, and this is one meaning of progress.

Ptolemy distinguished two kinds of astronomical prediction, both of which he regarded as valid. The first is the sort pursued in the *Mathematical Syntaxis* (also known as the *Almagest*), providing a mathematical model of the motions of the planets. His predecessors had used circular motions to describe the movements of astronomical bodies, even though the paths of the Moon, Sun, and other planets do not appear to be circular to observers.

Some had worked with combinations of different regular circular motions to account for seeming irregularities. The introduction of the epicycle and deferent allowed more complicated motions to be described, still using circles. A planet is described as travelling along a circular path—an epicycle—whose centre moves around the Earth on another circular path (a deferent). The combined

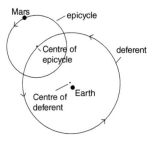

14. An epicycle and deferent for a planet.

motions allow for the appearance of changing speed and direction, even reversal (resulting in retrograde motion). Eccentric circles were also used, in which the centre of a circular path is displaced from the Earth. Ptolemy discussed the equivalence of epicyclic and eccentric models in accounting for the phenomena (Figure 14).

Before presenting his own mathematical model, Ptolemy outlined the working assumptions of his geocentric (Earth-centred) cosmos. Even his cosmology has a mathematical flavour: he explains that the heaven is spherical and moves as a sphere; to the senses, the Earth—taken as a whole—is spherical; the Earth is in the middle of the heavens, much like its (geometrical) centre; the Earth has the ratio of a point to the distance of the 'fixed' stars; the Earth has no motion from place to place. Ptolemy introduced another innovation to his mathematical model: the centre of an epicycle appears to sweep out equal angles in equal times, from an imagined point known as the equant. He accounted for planetary motion so well that the *Syntaxis* was the foundation of mathematical astronomy for many centuries (Figure 15).

Astrological prediction

The second type of astronomical prediction—which we call astrology—is the subject of Ptolemy's *Tetrabiblos*, and investigates

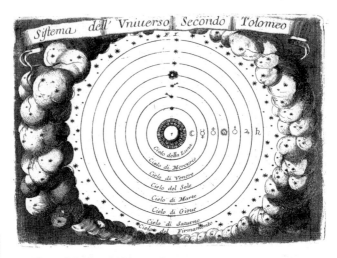

**15. The Ptolemaic world system from Vincenzo Maria Coronelli's
Atlante veneto (Venice, 1691).**

the changes that the motions of the heavenly bodies bring about
on Earth. Ptolemy assumed the influence of the celestial on the
terrestrial as, in his view, a relationship between the celestial and
the terrestrial occurs necessarily, and accounts for some events on
Earth, affecting weather, for example, and aspects of human life.
Both disciplines of astronomy were worth pursuing, even if
astrology was less certain in its results than mathematical
astronomy, on which it depended. Ptolemy acknowledged that in
his own time some people disparaged astrology. In certain cases,
the issue may not be astrology itself, but those who put themselves
forward as experts while lacking sufficient knowledge or expertise.

Two different branches of astrology were used to make
predictions. The first, more general, applied to different regions,
ethnic groups, and cities, and dealt with major events such
as wars, famines, earthquakes, seasons of the year, and
meteorological conditions. The second specifically related to

16. What may be a celestial globe is shown in this scene of a birth depicted on a Roman sarcophagus, probably from about 176–93 CE (possibly from the eastern Mediterranean region).

individuals, each of whom was also subject to the more general effects. Planets were held to have influence over people's lives, including what occupation they pursue (farming, teaching, thieving, etc.) and personal relationships, such as marrying someone older or younger, the temperament of one's spouse, and friendships (Figure 16).

Countering those who claimed that astrology speaks only about those things that will happen in any event—implying that there is no point in trying to do anything about those things—Ptolemy explains that knowing about events in advance diminishes the possibility of panic or untoward exhilaration, enabling us to be calm regardless of what happens. But he warned that we cannot always get astrology 'right' and make predictions absolutely accurately. Nevertheless, he suggested that astrology's lack of total reliability shouldn't be disparaged, drawing an analogy to the pilot of a ship, relying on navigation that is not always infallible. Knowledge of astrology gives us the means to live life more peacefully and calmly, thus providing us with practical means to improve the quality of our lives. (Here Ptolemy may have been influenced by the views of Hellenistic philosophers, Epicurean and/or Stoic.)

Aiming for success

In antiquity, astrology was regarded as a *technē*, a term used to describe various areas of human endeavour, from the exalted—such as philosophy and mathematics—to the more mundane, such as sandal-making. Some English definitions of *technē* (plural = *technai*) include 'skill', 'craft', and 'art'; the term 'applied science' is also sometimes used. Whether particular practices—such as medicine and rhetoric—qualified as *technai* was not universally agreed. Furthermore, it was not simply craft status that accounted for the lack of one-to-one correspondence between theoretical principles and practical success. Some *technai* share a special characteristic, which Ptolemy pointed to when he described astrology as a stochastic (from the Greek word for 'aim') *technē*, comparing it to medicine. A stochastic *technē* is one that aims to produce a particular effect, the success of which cannot be assured; some failure is to be expected because of the nature of the endeavour. The result cannot be guaranteed, as certain things are outside the control of the practitioner, in this case, the astrologer or physician. Indeed, Galen affirms the stochastic character of both diagnosis and prognosis with regard to disease and health.

Other stochastic *technai*, such as navigation and rhetoric, can also only aim at success. Lack of success is not necessarily due to a defect in the practitioner or in the body of knowledge employed. Failure in navigation or medicine contrasts with that which may occur in a non-stochastic *technē* such as sandal-making, where a botched pair of sandals would be regarded as the result of a lack of competence on the part of the cobbler. Educated guesswork or

Ancient Greek and Roman Science

conjecture was often used out of necessity by practitioners of stochastic *technai* in the face of inconclusive evidence, incomplete knowledge, and the impossibility of full control of all relevant factors. Many ancient *technai* were regarded as stochastic, or capable only of aiming at success. The ancients recognized that not all of what we moderns might call a science was exact; the uncertain nature of much ancient 'scientific' endeavour was acknowledged, rather than denigrated. Furthermore, the idea of the possibility of certain knowledge was by no means universally held. Medicine was regarded as a difficult and tentative practice, and astrology, agriculture, navigation, and military leadership were also seen to employ a degree of guesswork.

Ptolemy's view of astrology was pragmatic. Astrology—particularly concerning individuals—was conjectural, like medicine, because of the many variable factors to be taken into account: the race, country, upbringing, and so forth, of the person. Some events occur as the result of general circumstances, not those of the specific individual. Ptolemy pointed to many ways in which astrology is not foolproof, and is liable to fail to provide certain knowledge. For him, the admission of fallibility does not require the rejection of the practice. In his opinion, we should welcome what is possible and be content. Like some other ancient thinkers, including Epicurus, Ptolemy emphasized that in certain areas of human endeavour, knowledge cannot be guaranteed; rather, we can only aim at knowing. Nevertheless, he urges us to strive for more and better knowledge, as a means to understand the world around us and—more importantly—to bring ourselves into harmony with the world itself.

Ptolemy explained that he would attempt to accomplish his goal in the *Syntaxis* by learning those things that had already been mastered by others and striving to make as much advancement as possible. To some extent, his work would be based on that of his predecessors, but he was committed to making his own contribution, to mark his own progress. In his view, this is how one achieves the most benefit from theoretical philosophy. The idea of progress has Stoic overtones, for *prokopē* (progress) was one of their catchwords. But others, including some Platonists, used the term too. Significantly, in Stoic philosophy discussions of progress are in the context of ethics.

For Ptolemy, studying and teaching mathematical theories were ethical endeavours. Ethical virtues may be had without education, but the greatest virtue—knowledge of mathematical theory—is obtained only through education, which depends on teaching. Because mathematics is the best philosophy, by teaching mathematical theories Ptolemy demonstrated that he had not only obtained that greatest of virtues, he also enabled others to achieve it. But Ptolemy recognized that not everyone had his mathematical skills and knowledge. A number of his works were aimed at students, as well as those who might wish to take shortcuts in making mathematical calculations for use in both astrological and meteorological prediction and medical practice.

Ptolemy commended the Egyptians for bringing together astrology and medicine. Noting that Hippocrates had incorporated astronomy into his own practice, Galen, in his work entitled *The Best Doctor is also a Philosopher*, stressed the importance of knowledge of astronomy and geometry for physicians, in helping to predict the course of disease, particularly with regard to changes in weather and the effects on human health. Galen also produced a commentary on the Hippocratic work *Airs, Waters and Places*, one of the earliest Greek medical texts, which highlights the possible effects of the seasons and weather on health.

Astrology provides a means of predicting things of a general, including seasonal, nature as well as the more individual factors that may impact on health and disease. While Galen stressed the importance of astronomical knowledge for medical practitioners, he criticized those astrologers who claimed to be able to predict the whole of one's life using a horoscope. Galen himself had considerable knowledge of technical astronomy (as well as astrology), some of which he learned from his father, who was an architect and builder with wide-ranging interests and who facilitated Galen's education. Galen was proud of his father's achievements in geometry, practical calculation, and astronomy; it seems that excellence in mathematics was a family trait, going back generations. When he complains about some who claim to be able to make all sorts of predictions about individuals on the basis of astrology, he is concerned that they really don't know enough mathematics, including astronomy. For Galen, it was no surprise that Roman doctors knew so little, given that Roman astrologers themselves had scant knowledge. However, he recognized that some people had to take shortcuts, especially when they lacked sufficient detailed understanding; this might apply to astrologers as well as to doctors.

Galen's calculations

Galen wrote a book to help those physicians in need of astronomical information, *The Stars of Hippocrates and the Geometry Useful in the Science of Medicine*, to aid in determining equinoxes and solstices, and to be used in attempting to predict changes in health and disease. The Hippocratic doctrine of 'critical days' pointed to specific days—and in some cases, hours—in which the course of disease changes; the result could literally be the difference between life and death. Traditionally, physicians had deemed the most important 'crises' (turnings; critical points; judgements) to occur one, two, or three weeks after a patient took to their bed. To further complicate things, medical practitioners debated whether it was the 20th or 21st day that was actually critical, or both.

Galen addressed the question of which of these days is most significant in *On Critical Days* by setting the Hippocratic doctrine of medical crises on an astrological basis, stressing the Moon's importance in relation to the onset of medical events. He suggests that two different lunar effects should be taken into account: one based on the time taken to return to a certain position with respect to the fixed stars (the sidereal month, calculated as $27\frac{1}{3}$ days), the other on a complete cycle of phases of the Moon (the synodic month, taken to be $29\frac{1}{2}$ days). Reasoning that, on average, the (new) Moon is not visible for three days a month (at the time of conjunction with the Sun) and so its effects then are negligible, he subtracts those days from the synodic month, resulting in $26\frac{1}{2}$. He proposes a new 'medical month' ($26\frac{11}{12}$ days), the arithmetic mean of the sidereal ($27\frac{1}{3}$ days) and the shortened synodic month ($26\frac{1}{2}$ days), which can be divided into four medical weeks ('hebdomads', periods of seven). By his calculations the third hebdomad exceeds 20 days by only a few hours, therefore the 20th day should be regarded as more critical, in agreement with Hippocrates' teaching and the conclusion of Galen's third medical week. Emphasizing the advantages of his computations, Galen highlights the accuracy obtained by following his method for determining the day of crisis. While his disdain for the poor practices of some of his competitors is evident, he makes it clear that he can save the day, with a new astrologically informed calculation of the day of medical crisis.

Accurate knowledge of the anatomy of the body was also crucial. Galen stressed the importance of observation for the medical practitioner and produced a manual to instruct others, *Anatomical Procedures*. His dissections provided new information (including that urine flows only in one direction), and he changed his own views when confronted by new anatomical evidence (e.g. regarding the muscles in toes and fingers). His understanding of nerves was due to the detailed experiments he performed. Accuracy was key for Galen, and an important aspect of improvement and advancement in medical practice.

Galen also wrote on philosophical topics, reporting that as a young man he had extensively studied both philosophy and mathematics. Galen advised students of medicine to begin with physical theory—concerning the cosmos, matter, and living things—before reading his work *The Method of Healing*. Well versed in Peripatetic and Stoic logic, Galen argues for the necessity of using formal logic and demonstration in explaining medical ideas. For Galen, the best and most reliable kind of knowledge is achieved through 'scientific demonstration' (*epistēmonikē apodeixis*), following Aristotle's method of presenting demonstrations in syllogistic form. The premises used in demonstration must be propositions for which the truth is evident, via sense perception or reason. Galen appears to count those opinions held by the majority, or by consensus, as evident to reason, and traditional authority carries much weight for him. However, while he had the highest regard for some of his predecessors (especially Hippocrates and Plato) and used some of their ideas as a basis for his own arguments, he did not shrink from disagreeing with them on details. And, while students should have a good understanding of philosophical demonstration, they must recognize that knowledge itself is stochastic.

Galen thought it possible to make progress in scientific endeavours, particularly medicine, by using reasoning together with empirical evidence. He pointed to his own new discoveries of things previously unknown, for example as previously mentioned, regarding nerves. But he also claimed credit for discoveries outside medicine, such as the introduction of relational syllogism. Here is an example:

> Theon has twice as much as Dion, but Philon has twice as much as
> Theon; therefore, Philon has four times as much as Dion.

With this innovation, Galen thought he had improved on Aristotelian and Stoic logic; he regarded it also relevant for mathematics, including geometrical proofs.

Both Galen and Ptolemy believed it possible to make new contributions to knowledge, building on the work of predecessors and attempting to contribute as much advancement as possible. They aimed to do something new, to benefit those to come in the future. Given that their work was still being used more than 1,000 years later, it seems that they succeeded.

Chapter 11
Going by the book—or not

Late antiquity—the historical period involving the transition from antiquity to the medieval era in Europe and the Mediterranean, roughly the end of the 3rd to the 6th centuries CE—is notable for the significant number of commentaries produced focusing specifically on philosophical, scientific, and mathematical texts, including works by Hippocrates, Plato, Aristotle, and Ptolemy. Ancient commentators used the genre as a platform for communicating their own new ideas, fresh perspectives, innovative methods, and—sometimes radical—interpretations. Simplicius expresses his aim to make Aristotle's work accessible to others, but he does not shrink from giving Neoplatonic readings, emphasizing agreement between the ideas of Aristotle and Plato.

Later generations engaged with the work of earlier thinkers in various ways, adopting and adapting their ideas, quoting, and criticizing. The commentary was an important genre of scientific communication, particularly in later antiquity, and commentaries were produced on all sorts of texts. Commentators engaged critically with their 'target' texts (those on which they were commenting), the authors of those texts, and with other commentators as well. Commentators typically stress that they are explaining and clarifying earlier works; some 'correct' those works, to fit their own ideas and interpretations (Figure 17). Hipparchus,

17. *Alexander of Aphrodisias and Aristotle* by Andrea Briosco,
16th-century plaquette.

in his *Commentary on the Phainomena of Aratus and Eudoxus*,
pointed to errors in Aratus' work.

Mathematical texts were targeted by some commentators who
were not necessarily mathematicians themselves; Proclus'
5th-century CE commentary on the Euclidean *Elements* focuses

more on philosophical questions than on geometry. And while commentators may have presented themselves as explaining the ideas of others, their explanations did not always include agreement with those ideas. John Philoponus, writing in the 6th century CE, was generally friendly towards Aristotle in his commentaries; however, he launched an attack in his work *Against Aristotle on the Eternity of the World*. A Christian Neoplatonist, Philoponus repeatedly asserted the doctrine of creation, arguing against the existence of the fifth element, the *aither*, which is, according to Aristotle, eternal and divine. He rejected Aristotle's claim that the world is eternal because, if the world is eternal, creation is denied. Other commentators also disagreed with the authors of their target texts, though perhaps not so fundamentally. Furthermore, even though they were seemingly focused on the explication of texts, commentators brought new observations, data, and methods to bear in their elucidation of scientific ideas.

Commentators played an important role in shaping the way the works of their predecessors were read and regarded, and some texts survive only in the language into which they were translated by them. In the 4th century CE, Calcidius wrote a commentary on Plato's *Timaeus*; his Latin translation (to section 53c) represented Plato to Latin readers throughout the Middle Ages. In certain instances, commentators were keen to preserve texts that might otherwise have been lost. In his commentary on Archimedes' *Sphere and Cylinder*, Eutocius of Ascalon, writing in the 6th century CE, included 12 solutions to the problem of doubling the cube. It is due to Eutocius' commentary cum anthology that we have the text of Eratosthenes' *Letter to King Ptolemy*.

Not all commentaries were intended for the same type of audience; scientific topics interested people coming from diverse perspectives and wanting differing approaches. Furthermore, individual commentators had their own motivations and aims. Some commentaries were produced within educational settings,

The survival of ancient Greek texts in translation

A number of ancient Greek works survive only in translations into other languages, including Latin, Arabic, Syriac, and Hebrew. Examples include Galen's commentary on the Hippocratic *Airs, Waters, and Places*, which is lost in the original Greek, but survives in an Arabic translation, a Hebrew version, and a Latin translation derived from the Hebrew. Theophrastus' work on meteorology survived only, and partially, through Arabic and Syriac translations. Ptolemy's *Planetary Hypotheses* was originally two books, providing a physical description of the universe, including the planetary spheres, distances, sizes, and motions. In Greek, only the first part of Book 1 survives, and three 16th- and 17th-century Latin translations derive from the incomplete Greek. A Hebrew version also exists, but the complete work is extant only in Arabic. Without the later translations of these Greek works into other languages, they would be—in some cases—completely lost to us.

aimed at students; others were produced for more advanced readers. A number of important commentaries were produced by specialist astronomers and professional physicians. Galen produced as many as 17 commentaries on Hippocratic works. While he was much concerned with medical matters, Galen also had deeply philosophical interests, and discussed textual and philological matters of scholarship. Over time, Galen's writings themselves became the focus of commentaries by later authors, including the 6th-century CE physician Stephanus of Athens.

Specialist mathematical texts, including those by Apollonius of Perge (*fl.* 200 BCE) and Claudius Ptolemy, were also the subject of commentaries, written by authors with a high level of mathematical expertise, including Theon of Alexandria, at work in the second half of the 4th century CE, and his daughter

Hypatia, who died in 415 CE. Of the numerous commentaries produced on philosophical texts, many of which may be regarded as philosophical works in their own right, some were intended for publication (such as those by Alexander of Aphrodisias and Simplicius on works by Aristotle), while others were transcriptions of lectures. We have glimpses of the culture in which commentaries were produced and shared, providing evidence of well-stocked libraries. Porphyry, in his *Life of Plotinus* composed in the 3rd century CE, makes it clear that texts were read aloud, and as a group activity. Philoponus divided his commentaries into portions that could be delivered orally within the course of an hour.

Women and mathematics

Mathematics was a male-dominated activity in antiquity. Nevertheless, there is evidence of women making original contributions at a very high level, as well as being involved in teaching. Music was an important branch of mathematics for the ancient Greeks, and thanks to Porphyry, in his *Commentary on Ptolemy's Harmonics*, we have some information about Ptolemaïs of Cyrene, the only Greek female music theorist known to us, active perhaps in the early 1st century CE. Her work, *The Pythagorean Elements of Music*, with its question-and-answer format, may have been a teaching text.

Pandrosion, a mathematics teacher, is the dedicatee of the third book of Pappus of Alexandria's 4th-century CE *Collection*, in which he considers solutions to various mathematical problems. Pappus deviates from the usual practice of being complimentary in a dedication, opining that Pandrosion may not have been the most effective instructor, and claiming that some of her self-proclaimed students were incompetent. While he is critical

(continued)

of the efforts of some of her students, this may be due to some sort of rivalry. No writings by Pandrosion survive, nor do we have other mentions of her or her work. It is impossible to judge Pappus' criticisms.

Hypatia was a teacher of mathematics and Neoplatonist philosophy in Alexandria, having studied with her father Theon and others. She wrote on various topics in mathematics, astronomy, and philosophy, including planetary motions and number theory; she also edited works written by others. Synesius of Cyrene, Bishop of Ptolemais, praised her excellence as a teacher, explaining that she had taught him everything that he knew about the astronomical instruments known as astrolabes. Murdered by Christians, Hypatia is often represented as a pagan martyr.

John Philoponus

Most ancient commentaries on philosophical works, including those concerned with the physical world, are organized by *lemmata* (plural of *lemma*), that is, quotations of words or phrases from the 'target' text which is being commented upon. In his commentary on Aristotle's *Meteorology*, Philoponus quotes the opening passage of the Aristotelian text:

> The first causes of nature, all natural motion, also the stars disposed in order in the celestial revolution, the corporeal elements, their number and quality, and their transformation into each other, as well as coming-to-be and perishing in general have been dealt with before.

He then comments:

> Immediately at the beginning Aristotle lists his own works on nature, all those that have preceded the *Meteorology* in the natural order, giving the first place to the so-called *Course on Physics*, to

which the common name of *Physics* was assigned as its own proper title because it treats of the common principles of all natural phenomena, for which reason it also precedes the others.

Philoponus is sharing his knowledge and understanding of Aristotle's writings. However, as we have seen, he did not always agree with Aristotle, and was concerned to distinguish his own views, even as he sought to explicate Aristotle's.

Philoponus frequently presented his own ideas, in his own way. On occasion, this involved the use of mathematics. We find him invoking geometrical language and a lettered diagram in commenting on a passage in Aristotle's writings in which Aristotle had not himself done so. Philoponus also produced a commentary on Nicomachus' *Introduction to Arithmetic*. There is other evidence of his strong interest in mathematics, including astronomy; a work on the astrolabe has been ascribed to him (Figure 18). His interest and expertise in mathematics coloured his explication of Aristotle's views.

For example, Philoponus explains that Aristotle was concerned to explain shooting stars and related phenomena as things that are moving. He then describes a geometrical shape (a rectangle) and uses geometrical terminology (the diagonal) to set up an analogy to a presumably familiar experience—or what may even be a thought experiment—of ants tracking a path (Figure 19):

> Assume a rectangle ABCD, with its cross-section, i.e. diagonal, AD; let two ants of equal strength be moving, one from C to A and another, again, from B to A. When they are at A and neither gets the better of the other as they push each other, they are shoved off the sides of the rectangle, and being deflected, they get carried off along the cross-section AD.

Philoponus' use of an analogy to shared everyday experience is an explanatory tactic often employed by Aristotle; in fact, Aristotle

18. Planispheric astrolabe, 14th century, probably English.

drew an analogy in his own discussion of these phenomena to the motion of thrown objects. But when Philoponus invites us to 'let two ants of equal strength be moving', describing their motion on a plane, he invokes the Euclidean language of geometrical proof. He also referred to a lettered diagram, another hallmark of geometry. Aristotle had earlier referred to lettered diagrams in his discussions of different winds and the rainbow, but he did not refer to any diagram in his explanation of shooting stars; nor did the other commentators on the same Aristotelian text.

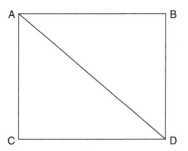

19. Lettered diagram illustrating Philoponus' description of the path tracked by moving ants.

Philoponus proved he was an innovator in his discussion of shooting stars. He employed a technique found in Aristotle and utilized by other commentators, the use of geometrical diagrams to explain physical phenomena. But Philoponus found a new application, explaining a different phenomenon. And, perhaps most surprisingly, while Philoponus' explanation is cast in geometrical terms, it is not really a mathematical argument. The rather unexpected introduction of imaginary ants is a striking example of Philoponus using Aristotle's explanatory tactics (here, both analogy and diagrams) to support a novel and vivid explanation. The analogy between the motion of shooting stars and the motion of ants emphasizes that these phenomena occur in the same sort of physical space.

The individual voice of the commentator resounded through the genre, even as, in late antiquity particularly, complex relationships developed between Graeco-Roman culture and the new religions of Christianity and, later, Islam. Commentary continued to be a very important genre in the medieval period, used in different cultures and contexts, Christian, Jewish, and Islamic, to reflect and expand on many types of writings. The survival or loss of ancient works, including those on scientific topics, in many cases depended on the attitudes of individual Christians and Muslims. The preservation of pagan texts relied, in large part, on their perceived value within the ascending cultures of Christianity and Islam, and to some extent that of Judaism, with its tradition of rabbinic commentary. Later commentators on ancient scientific, medical, and mathematical works—in different linguistic, religious, and cultural traditions—incorporated and referred to the work of the ancient commentators, allowing the voices of distant and foreign predecessors to continue to be heard, and to be reckoned with afresh.

Chapter 12
Beyond antiquity

The legacies of ancient Greek and Roman science—including ideas, practices, instruments, and ways of communicating—endured through the Middle Ages, Renaissance, and even into the modern period. A number of concepts and concerns are seemingly very long-lived, including those relating to elements, atoms, chaos, the idea of a microcosm/macrocosm, the question of whether the cosmos is itself a living being, the status of scientific knowledge in wider culture, the place of mathematics in science, and whether the world changes in significant ways over time. This apparent longevity points to a degree of continuity among scientific practitioners and across communities. While there were many discontinuities, the engagement with similar questions in different contexts at distinct times and places may indicate some shared interests and ambitions, expressed as a sort of scientific community—or communities—extending over cultural, temporal, and linguistic boundaries.

The influence of ancient thinkers—including Pythagoras, Plato, Aristotle, Epicurus, and the Stoics—remained strong throughout the medieval and early modern periods, and shaped how philosophy, science, and mathematics were understood and pursued. The division of labour between mathematicians and philosophers continued in formalized ways, through the organization of education. The universities that were founded in the Middle Ages followed an Aristotelian curriculum, and this prevailed throughout the Renaissance, even into the 18th century in some places. Arithmetic, harmonics, geometry, and astronomy became the medieval quadrivium, necessary for study in university. In the early modern period, the distinction between the proper role of mathematicians (including astronomers) versus that of philosophers (including natural philosophers) was debated, even as there were shifts in disciplinary lines.

Some modern scientists feel a powerful affinity with their ancient intellectual forebears. While Anaximander's understanding of the *apeiron* may be hard to grasp, some 20th-century physicists found it a helpful concept. In the 1930s and 1940s, physicists worked to turn quantum mechanics (a theory of particles and their interactions) into a theory of particles and fields (i.e. quantum field theory). But they encountered difficulties, as attempts to do so often led to the appearance of infinities in their equations—indicating that they were not representative of physical possibilities. Attempting to dodge this problem, in 1944, Max Born and H. W. Peng, working in the Department of Mathematical Physics at the University of Edinburgh, published a letter to the editors of the highly respected journal *Nature*, on the 'Statistical Mechanics of Fields and the "Apeiron"', suggesting that the word *apeiron*, introduced by Anaximander to describe the 'boundless and structureless primordial matter', be used to refer to the subgroup of pure states that is intermediate between the ordinary notion of a particle or quantum and that of a mechanical system. The ancient and modern uses of *apeiron* are very different. Even though their suggestion did not take hold, at a time of crisis of explanation Born and Peng looked back to ancient Greek ideas for inspiration. Born was awarded the Nobel Prize in Physics in 1954.

Another awardee of the Nobel Prize in Physics (1921), Albert Einstein, traced the roots of modern science to ancient Greece. He argued that a scientist 'may even appear as "Platonist" or "Pythagorean" insofar as he considers the viewpoint of logical simplicity as an indispensable and effective tool of his research'. The English mathematician and philosopher Alfred North Whitehead, in his 1925 Lowell Lectures on 'Science in the Modern World', credited Pythagoras with founding European philosophy and mathematics.

The origin and early history of scientific ideas and practices of ancient Greeks and Romans have also fascinated philosophers. In 1841, Karl Marx completed his doctoral thesis in philosophy, on 'The Difference between the Democritean and Epicurean Philosophy of Nature'. Friedrich Nietzsche, who trained as a philologist and had worked on the dating of *The Contest of Homer and Hesiod*, delivered a series of lectures at the University of Basel between 1872 and 1876, in which he surveyed the Greek philosophers from Thales to Socrates.

Modern, and ancient, science

In the mid-20th century the Nobel prizewinning physicists Werner Heisenberg (1932 prize) and Erwin Schrödinger (1933) both published books on early Greek science, intended for non-specialists. They were each important contributors to the new quantum theory of the early 20th century, but their formulations were very different, and neither approved of the other's, even though these turned out to be—with some qualification—equivalent. And they had much else in common. They each had an education that included the ancient classics, during which they were exposed to ideas of the ancient Greeks. By his own account, Schrödinger enjoyed the study of classical languages when he was young, and that interest seems to have remained throughout his life. Heisenberg's father was a professor of Greek (not only ancient) philology in Munich. Heisenberg read Plato's *Timaeus* in Greek in secondary school. Born and Einstein would also have had a similar education. These physicists were interested, like the ancient

Greek philosophers themselves, in more general—as well as ethical—philosophical issues.

In 1948 Schrödinger gave four public lectures at both University College Dublin and University College London on the subject of 'Nature and the Greeks'. He developed his lectures into the book that was published in 1954. Heisenberg published his *Physics and Philosophy*, derived from his 1955–6 Gifford Lectures at the University of St Andrews, in 1958. They both engaged with the work of ancient Greek scientific thinkers, comparing and contrasting the new physics of the 20th century to that of ancient Greece and Rome.

Heisenberg points to the view attributed to Thales, that water is the material cause of everything, noting approvingly that Nietzsche claimed that it expresses three fundamental ideas of philosophy: the question of the material cause of things; the demand that this question be answered in conformity with reason, without resorting to myths; and the idea that—ultimately—it must be possible to reduce everything to one principle.

Heisenberg emphasized that the use of experimentation and mathematics, which he dated to Galileo and Newton, is what distinguishes modern science from what came earlier, in his view providing modern science with a firmer basis. Nevertheless, he concludes that some claims of ancient philosophy are very near to those of modern science. For Heisenberg, this shows how far it is possible to get by combining ordinary experience of nature with an effort to order this experience logically, aiming to understand it from first principles. Aristotle probably would have found much to agree with here. While pointing to a fundamental difference between modern and ancient science—an emphasis on experiment and replicability—Heisenberg highlighted the shared desire to find general principles to explain nature.

Schrödinger sought to show that the fundamental features of the modern scientific world-picture are historically produced, rather

126

Greek philosophers themselves, in more general—as well as ethical—philosophical issues.

than logically necessitated, pointing to two features of modern science that he thought could be traced back to the Presocratics. He credits the Milesians with the idea that the world can be understood. The second feature of modern science, objectivation, Schrödinger found also in Heraclitus' approach: in order to understand the world, Heraclitus became an external observer.

Schrödinger highlighted what he took to be a competition among ancient Greeks between the emphasis on the application of reason versus the valorization of sensory experience. Democritus received special commendation for having an advanced epistemology, in which understanding involves sense perception. In praising Democritus, Schrödinger quoted a fragment preserved by Galen and published by Hermann Diels (1848–1922), in *The Fragments of the Presocratics* (*Die Fragmente der Vorsokratiker*, 1903), in what he described as 'the famous dialogue between the intellect and the senses':

Beyond antiquity

> (Intellect:) Sweet is by convention and bitter by convention, hot by convention, cold by convention, colour by convention; in truth there are but atoms and the void.

> (The Senses:) Wretched mind, from us you are taking the evidence by which you would overthrow us? Your victory is your own fall.

Schrödinger's interest in those early Greeks who sought to understand nature—shared with other 20th-century thinkers—was fuelled by Diels' publication. Diels was a German classicist and historian with a special interest in ancient science and technology. His schoolboy hobby had been conducting chemistry experiments. His final work was a translation of Lucretius' poem, published posthumously in 1924, with an Introduction by Einstein, praising his clarity.

When we refer to the early Greek philosophers as 'Presocratic', we follow Diels' definition. His edition presented the fragments of the Presocratic philosophers in a collection that was easy to access, organized by each individual thinker. His work was so well received that he published a new edition in 1906–10, followed by two more before his death; newer editions have since appeared. Diels greatly encouraged and facilitated the study of ancient Greek science, not only through the edition of the *Fragments*. And, because so much of what survives of the Presocratic philosophers is concerned with nature, these thinkers may appear to be closer to the modern ideal of the scientist than other philosophers, including Socrates, Plato, and even Aristotle. Today, we are still searching for answers to some of the same questions as the ancients: how did the world begin? What is the world made of? Are there other worlds? How do humans relate to other parts of the universe, including other living beings? How do we know what we think we know?

Note on dates and spelling

Whenever possible, dates provided in this *Very Short Introduction* are based on those in the *Oxford Classical Dictionary*, 4th edition.

For the most part (but not always), I have adopted a 'latinized' spelling of Greek names and terms, to conform to general usage.

References

Chapter 1: Understanding the world

Seneca, *Of Consolation: To Helvia*, section 20, in L. Annaeus Seneca, *Minor Dialogues Together with the Dialogue 'On Clemency'*, translated by Aubrey Stewart (London: George Bell and Sons, 1889).

Chapter 2: Expert poets

Homer, *The Odyssey*: annual cycle of recurrent seasons 'with the year's full circling the seasons returned', Book 11, line 295; translated by Anthony Verity (Oxford: Oxford University Press, 2016), p. 146.

Navigational advice: Homer, *The Odyssey*, Book 5, lines 269–77, translated by Anthony Verity (Oxford: Oxford University Press, 2016), pp. 68–9.

Hesiod, *Works and Days*, lines 609–16, and 619–23, translated by M. L. West (Oxford: Oxford University Press, 1988), p. 55.

Leaving room for materialist explanations: David Konstan, 'What is Greek about Greek Mythology?', *Kernos* 4 (1991): 11–30, p. 21. http://journals.openedition.org/kernos/280.

Chapter 4: Those clever Greeks

Aristophanes, *Clouds*, lines 227–34, translated by Stephen Halliwell (Oxford: Oxford University Press, 2015), p. 29.

Herodotus, *The Histories*, Book 2, sections 19 and 23, translated by Robin Waterfield (Oxford: Oxford University Press, 1998), pp. 102–3.

Plato, *Phaedo*, 96a–c, translated by David Gallop (Oxford: Oxford University Press, 1993), pp. 51–2.

Chapter 5: Let no one unskilled in geometry enter

Plato, *Timaeus*, 27a and 29c–d, translated by Robin Waterfield (Oxford: Oxford University Press, 2008), pp. 15 and 18.

Kepler's letter to Galileo, 13 October 1597, in *Johannes Kepler: Life and Letters*, translated by Carola Baumgardt (New York, NY: Philosophical Library, 1951), pp. 40–1.

Chapter 7: Old school ties

Ancient Greek and Roman Science

Epicurus on animal signs: Diogenes Laertius, *Lives of the Eminent Philosophers*, Book 10, sections 115–16, translated by Pamela Mensch (Oxford: Oxford University Press, 2020), p. 385.

Plunging into myth: Diogenes Laertius, *Lives of the Eminent Philosophers*, Book 10, section 87, translated by Pamela Mensch (Oxford: Oxford University Press, 2020), p. 380.

Chapter 8: Roman nature

Virgil, *Georgics*, Book 1, lines 204–11, translated by Peter Fallon (Oxford: Oxford University Press, 2006), pp. 12–13.

Pliny on Virgil: *Natural History*, Book 18, chapter 56, translated by John Bostock (London: Taylor and Francis, 1855).

In this blessed peace: Pliny, *Natural History*, Book 2, chapter 45, translated by John Bostock (London: Taylor and Francis, 1855).

Chapter 9: River deep, mountain high

Lynn Thorndike, Query 7 in 'Notes and Correspondence', *Isis* 9, no. 3 (1927): 425–6.

On Xenagoras: Plutarch, *Aemilius Paullus*, section 15, in *Roman Lives*, translated by Robin Waterfield (Oxford: Oxford University Press, 1999), p. 54.

Archimedes, *The Sand-Reckoner*, in *The Works of Archimedes*, translated by Thomas L. Heath (New York, NY: Dover

Publications, by arrangement with Cambridge University Press, 1912; reissue of 1897 edition of Heath), pp. 221–32, on pp. 221–2.

Cattle problem opening: in *Selections Illustrating the History of Greek Mathematics*, vol. 2: *From Aristarchus to Pappus*, translated by Ivor Thomas (with slight emendation) (Cambridge, MA: Harvard University Press, 1941), p. 203.

Chapter 10: Is there scientific progress?

Galen's medical weeks: Stephan Heilen, 'Galen's Computation of Medical Weeks: Textual Emendations, Interpretation History, Rhetorical and Mathematical Examinations', *SCIAMVS: Sources and Commentaries in the Exact Sciences* 19 (2018): 205–6, 211.

Galen's relational logic: Galen's *Institutio Logica*, translated by John Spangler Kieffer (with slight emendation) (Baltimore, MD: Johns Hopkins University Press, 1964), p. 49.

Chapter 11: Going by the book—or not

Philoponus, opening passage, Book 1, chapter 1: *Philoponus On Aristotle Meteorology 1.1–3*, translated by Inna Kupreeva (London: Bristol Classical Press, 2011), p. 31.

Philoponus on ants tracking a path, Book 1, chapter 4: *Philoponus On Aristotle Meteorology 1.4–9, 12*, translated by Inna Kupreeva (London: Bristol Classical Press, 2012), p. 48.

Chapter 12: Beyond antiquity

Max Born and H. W. Peng, 'Statistical Mechanics of Fields and the "Apeiron"', *Nature* 153 (1944): 164–5.

Einstein: for the complete passage, see 'Reply to Criticisms', in *Albert Einstein: Philosopher-Scientist*, ed. Paul Arthur Schilpp (Evanston, IL: The Library of Living Philosophers, 1949), p. 684.

On Democritus: Erwin Schrödinger, *Nature and the Greeks* (Cambridge: Cambridge University Press, 1954), pp. 29–30, 87, following the translation of fragment 125 by Cyril Bailey, *The Greek Atomists and Epicurus: A Study* (Oxford: Oxford University Press, 1928), pp. 178–9.

Further reading

Ancient texts

Those written texts that survive serve as our principal sources of information about science, medicine, and mathematics in ancient Greece and Rome. In many cases we have excellent editions that are easily accessible, for example, often in the Loeb Classical Library series, available online (<https://www.loebclassics.com/>); other valuable digital resources include the Perseus Digital Library (<http://www.perseus.tufts.edu>, available without charge).

Collections of selections from primary sources in translation include: Morris R. Cohen and I. E. Drabkin, *Source Book in Greek Science* (Cambridge, MA: Harvard University Press, 1948); Georgia L. Irby-Massie and Paul T. Keyser, *Greek Science of the Hellenistic Era: A Sourcebook* (London: Routledge, 2002); *Selections Illustrating the History of Greek Mathematics*, 2 vols, ed. Ivor Thomas (Cambridge, MA: Harvard University Press, 1939–41).

Material evidence

Alexander Jones, ed., *Time and Cosmos in Greco-Roman Antiquity* (Princeton, NJ: Princeton University Press, 2016); James Evans, 'The Material Culture of Greek Astronomy', *Journal for the History of Astronomy* 30 (1999): 237–307; Richard J. A. Talbert, *Roman Portable Sundials: The Empire in your Hand* (Oxford: Oxford University Press, 2017); Anthony J. Turner, *Mathematical*

Instruments in Antiquity and the Middle Ages: An Introduction (London: Vade-mecum, 1994); on the Antikythera Mechanism: Alexander Jones, *A Portable Cosmos: Revealing the Antikythera Mechanism, Scientific Wonder of the Ancient World* (New York: Oxford University Press, 2017); Patricia A. Baker, *The Archaeology of Medicine in the Greco-Roman World* (Cambridge: Cambridge University Press, 2013); and Ralph Jackson, 'Roman Doctors and their Instruments: Recent Research into Ancient Practice', *Journal of Roman Archaeology* 3 (1990): 5–27.

We are now fortunate to have virtual access to objects and information in various collections and from many archaeological sites. The following websites are only a small sample of what is available, to view examples of vase paintings, mosaics, coins, and other ancient artefacts relevant for the study of ancient science: the Ancient Sundials research project in Berlin surveyed and analysed all known Greek and Roman sundials (<http://repository.edition-topoi.org/collection/BSDP/>); Surgical Instruments from Ancient Rome: A Display of Surgical Instruments from Antiquity, Historical Collections & Services of the Health Sciences Library, University of Virginia (<http://exhibits.hsl.virginia.edu/romansurgical/>); British Museum (<https://www.britishmuseum.org/collection>); Metropolitan Museum of Art (<https://www.metmuseum.org/art/collection>); American Numismatic Society (<http://numismatics.org/search/>).

Encyclopedic resources

Antony Spawforth, Esther Eidinow, and Simon Hornblower, eds, *The Oxford Classical Dictionary*, 4th edn (Oxford: Oxford University Press, 2012, and <https://www.oxfordreference.com/view/10.1093/acref/9780199545568.001.0001/acref-9780199545568>); Edward N. Zalta, ed., *Stanford Encyclopedia of Philosophy* (<https://plato.stanford.edu/>); Paul T. Keyser and Georgia L. Irby-Massie, eds, *The Encyclopedia of Ancient Natural Scientists: The Greek Tradition and Its Many Heirs* (London: Routledge, 2008).

General introductions, handbooks, and histories

Alexander Jones and Liba Taub, eds, *The Cambridge History of Science*, vol. 1: *Ancient Science* (Cambridge: Cambridge University

Press, 2018); Liba Taub, ed., *The Cambridge Companion to Ancient Greek and Roman Science* (Cambridge: Cambridge University Press, 2020); Paul Keyser and John Scarborough, eds, *Oxford Handbook of Science and Medicine in the Classical World* (New York, NY: Oxford University Press, 2018); Georgia L. Irby, ed., *A Companion to Science, Technology, and Medicine in Ancient Greece and Rome—Blackwell Companions to the Ancient World*, 2 vols (Hoboken, NJ: John Wiley & Sons, 2016); T. E. Rihll and C. J. Tuplin, eds, *Science and Mathematics in Ancient Greek Culture* (Oxford: Oxford University Press, 2002); T. E. Rihll, *Greek Science* (Oxford: Oxford University Press for the Classical Association, 1999).

On ancient scientific, mathematic, and medical texts

Gian Biagio Conte, *Genres and Reader: Lucretius, Love Elegy, Pliny's Encyclopedia*, trans. Glenn W. Most (Baltimore, MD: Johns Hopkins University Press, 1994); Markus Asper, ed., *Writing Science: Medical and Mathematical Authorship in Ancient Greece* (Berlin: De Gruyter, 2013); Liba Taub, *Science Writing in Greco-Roman Antiquity* (Cambridge: Cambridge University Press, 2017).

Chapter 1: Understanding the world

Readers may wish to consult the sections of the *Cambridge History of Science*, vol. 1: *Ancient Science* on Mesopotamia, Egypt, India, and China; and also Francesca Rochberg, *Before Nature: Cuneiform Knowledge and the History of Science* (Chicago, IL: University of Chicago Press, 2016); and G. E. R. Lloyd, *Expanding Horizons in the History of Science: The Comparative Approach* (Cambridge: Cambridge University Press, 2021).

Chapter 2: Expert poets

For *The Contest of Homer and Hesiod*, see *Hesiod, the Homeric Hymns and Homerica*, trans. Hugh G. Evelyn-White (London: Heineman, 1914), in the public domain at <https://www.gutenberg.org/files/348/348-h/348-h.htm#chap88>. As an introduction to Hesiod, see Robert Lamberton, *Hesiod* (New Haven, CT: Yale University Press, 1988). Samuel Butler, *The Authoress of the*

Odyssey: Where and when she wrote, who she was, the use she made of the Iliad, and how the poem grew under her hands (London: Longmans, Green, 1897) is in the public domain at <https://www.gutenberg.org/ebooks/49324>.

See also D. R. Dicks, *Early Greek Astronomy to Aristotle* (Ithaca, NY: Cornell University Press, 1970), pp. 27–38, on Homer and Hesiod; and Robert Hannah, *Time in Antiquity* (London: Routledge, 2009).

Chapter 3: Inventing nature

On Thales and other Presocratic philosophers, G. S. Kirk, J. E. Raven, and M. Schofield (often referred to as 'KRS'), *The Presocratic Philosophers: A Critical History with a Selection of Texts*, 2nd edn (Cambridge: Cambridge University Press, 1983) remains an excellent starting point, for the fragments as well as commentary; see also Daniel W. Graham, trans. and ed., *The Texts of Early Greek Philosophy: The Complete Fragments and Selected Testimonies of the Major Presocratics*, 2 vols (Cambridge: Cambridge University Press, 2010); James Warren, *Presocratics* (Stocksfield: Acumen, 2007); *Lives of the Eminent Philosophers: Diogenes Laertius*, trans. Pamela Mensch and ed. James Miller (New York, NY: Oxford University Press, 2018); Alexander P. D. Mourelatos, *The Route of Parmenides*, rev. edn (Las Vegas, NV: Parmenides, 2008). See also James Warren, 'Diogenes Laërtius, Biographer of Philosophy', in Jason König and Tim Whitmarsh (eds), *Ordering Knowledge in the Roman Empire* (Cambridge: Cambridge University Press, 2007), pp. 133–49.

Chapter 4: Those clever Greeks

Rosalind Thomas, *Herodotus in Context: Ethnography, Science and the Art of Persuasion* (Cambridge: Cambridge University, 2000); K. J. Dover, *Aristophanes: Clouds* (Oxford: Clarendon Press, 1968), pp. xxxii–lvii, on Socrates; Brooke Holmes, *The Symptom and the Subject: The Emergence of the Physical Body in Ancient Greece* (Princeton, NJ: Princeton University Press, 2010); G. E. R. Lloyd, *Methods and Problems in Greek Science: Selected Papers* (Cambridge: Cambridge University Press, 1991).

Chapter 5: Let no one unskilled in geometry enter

On Pythagoras and the Pythagoreans: Charles H. Kahn, *Pythagoras and the Pythagoreans: A Brief History* (Indianapolis, IN: Hackett Publishing, 2001); Carl A. Huffman, ed., *A History of Pythagoreanism* (Cambridge: Cambridge University Press, 2014); Walter Burkert, *Lore and Science in Ancient Pythagoreanism*, trans. Edwin L. Minar (Cambridge, MA: Harvard University Press, 1972).

The literature on Plato is vast. One might begin by reading his *Timaeus*, and Gregory Vlastos, *Plato's Universe* (Oxford: Clarendon Press, 1975); Thomas Kjeller Johansen, *Plato's Natural Philosophy: A Study of the* Timaeus-Critias (New York, NY: Cambridge University Press, 2004); Sarah Broadie, *Nature and Divinity in Plato's Timaeus* (Cambridge: Cambridge University Press, 2011).

On mathematics: Otto Neugebauer, *Exact Sciences in Antiquity*, 2nd edn (Providence, RI: Brown University Press, 1957); Serafina Cuomo, *Ancient Mathematics* (London: Routledge, 2001); Reviel Netz, *The Shaping of Deduction in Greek Mathematics: A Study in Cognitive History* (Cambridge: Cambridge University Press, 1999); Wilbur Richard Knorr, *The Ancient Tradition of Geometric Problems* (Boston, MA: Birkhäuser, 1986); Andrew Barker, *The Science of Harmonics in Classical Greece* (Cambridge: Cambridge University Press, 2007); David Fowler, *The Mathematics of Plato's Academy: A New Reconstruction*, 2nd edn (Oxford: Clarendon Press, 1999), including pp. 199–204, on the inscription.

See also Emma Gee, *Aratus and the Astronomical Tradition* (Oxford: Oxford University Press, 2013).

Chapter 6: A theory of everything

On Aristotle: Jonathan Barnes, *Aristotle: A Very Short Introduction* (Oxford: Oxford University Press, 2000); Andrea Falcon, *Aristotle and the Science of Nature: Unity without Uniformity* (Cambridge: Cambridge University Press, 2005); Armand Marie Leroi, *The Lagoon: How Aristotle Invented Science* (London: Bloomsbury, 2014); Daryn Lehoux, *Creatures Born of Mud and Slime: The Wonder and Complexity of Spontaneous Generation* (Baltimore,

MD: Johns Hopkins University Press, 2017); Sophia Connell, ed., *The Cambridge Companion to Aristotle's Biology* (Cambridge: Cambridge University Press, 2021).

Theophrastus and the Peripatetics: William W. Fortenbaugh, Pamela M. Huby, and Anthony A. Long, eds, *Theophrastus of Eresus: On His Life and Work* (New Brunswick, NJ: Transaction, 1985); Gavin Hardy and Laurence Totelin, *Ancient Botany* (London: Routledge, 2015); Liba Taub, *Ancient Meteorology* (London: Routledge, 2003); *Aristotle: Problems*, 2 vols, trans. Robert Mayhew (Cambridge, MA: Harvard University Press, 2011); Pieter De Leemans and Michèle Goyens, eds, *Aristotle's* Problemata *in Different Times and Tongues* (Leuven: Leuven University Press, 2006); Robert Mayhew, ed., *The Aristotelian* Problemata physica*: Philosophical and Scientific Investigations* (Leiden: Brill, 2015); Leonid Zhmud, *The Origin of the History of Science in Classical Antiquity*, trans. Alexander Chernoglazov (Berlin: De Gruyter, 2008).

Chapter 7: Old school ties

D. N. Sedley and A. A. Long, *The Hellenistic Philosophers: Greek and Latin Texts*, 2 vols (Cambridge: Cambridge University Press, 1987); R. W. Sharples, *Stoics, Epicureans and Sceptics: An Introduction to Hellenistic Philosophy* (London: Routledge, 1996); Elizabeth Asmis, *Epicurus' Scientific Method* (Ithaca, NY: Cornell University Press, 1984); James Warren, ed., *Cambridge Companion to Epicureanism* (Cambridge: Cambridge University Press, 2009); David N. Sedley, *Lucretius and the Transformation of Greek Wisdom* (Cambridge: Cambridge University Press, 1998); Brad Inwood, *Stoicism: A Very Short Introduction* (Oxford: Oxford University Press, 2018).

Chapter 8: Roman nature

Seneca the Younger: Lucius Annaeus Seneca, *Natural Questions*, trans. Harry M. Hine (Chicago, IL: University of Chicago Press, 2010); Gareth Williams, *The Cosmic Viewpoint: A Study of Seneca's* Natural Questions (New York, NY: Oxford University Press, 2012).

Pliny the Elder: Trevor Murphy, *Pliny the Elder's* Natural History*: The Empire in the Encyclopedia* (Oxford: Oxford University Press,

2004); Aude Doody, *Pliny's Encyclopedia: The Reception of the Natural History* (Cambridge: Cambridge University Press, 2010); Roger French, *Ancient Natural History: Histories of Nature* (London: Routledge, 1994); Mary Beagon, *Roman Nature: The Thought of Pliny the Elder* (Oxford: Clarendon Press, 1992).

Poetry: Katharina Volk, *Manilius and His Intellectual Background* (Oxford: Oxford University Press, 2009); Steven J. Green and Katharina Volk, eds, *Forgotten Stars: Rediscovering Manilius' Astronomica* (Oxford: Oxford University Press, 2011); Liba Taub, *Aetna and the Moon: Explaining Nature in Ancient Greece and Rome* (Corvallis, OR: Oregon State University Press, 2008).

Chapter 9: River deep, mountain high

On measurement: Florian Cajori, 'History of Determinations of the Heights of Mountains', *Isis* 12, no. 3 (1929): 482–514; O. A. W. Dilke, *Mathematics and Measurement*, 2nd edn (London: British Museum, 1989); Richard Talbert and Kai Brodersen, eds, *Space in the Roman World: Its Perception and Presentation* (Münster: Lit Verlag, 2004); Daniela Dueck and Kai Brodersen, *Geography in Classical Antiquity* (Cambridge: Cambridge University Press, 2012); Donald Engels, 'The Length of Eratosthenes' Stade', *The American Journal of Philology* 106, no. 3 (1985): 298–311.

On mathematics: Reviel Netz, *Ludic Proof: Greek Mathematics and the Alexandrian Aesthetic* (Cambridge: Cambridge University Press, 2009); Mary Jaeger, *Archimedes and the Roman Imagination* (Ann Arbor, MI: University of Michigan Press, 2008); Max Leventhal, 'Counting on Epic: Mathematical Poetry and Homeric Epic in Archimedes' *Cattle Problem*', *Ramus* 44, nos 1–2 (2015): 200–21; Max Leventhal, *Poetry and Number in Graeco-Roman Antiquity* (Cambridge: Cambridge University Press, 2022); Maria Dzielska, *Hypatia of Alexandria*, trans. F. Lyra (Cambridge, MA: Harvard University Press, 1995); Edward J. Watts, *Hypatia: The Life and Legend of an Ancient Philosopher* (New York, NY: Oxford University Press, 2017); Alan Cameron, 'Isidore of Miletus and Hypatia: On the Editing of Mathematical Texts', *Greek, Roman and Byzantine Studies* 31, no. 1 (2005): 103–27.

Chapter 10: Is there scientific progress?

Ptolemy: *Ptolemy's Almagest*, trans. G. Toomer (London: Duckworth, 1984); Olaf Pedersen, *A Survey of the Almagest*, rev. edn by Alexander Jones (New York, NY: Springer, 2011); James Evans, *The History and Practice of Ancient Astronomy* (New York, NY: Oxford University Press, 1998); Gerd Graßhoff, *The History of Ptolemy's Star Catalogue* (New York, NY: Springer,1990); Liba Taub, *Ptolemy's Universe: The Natural Philosophical and Ethical Foundations of His Astronomy* (Chicago, IL: Open Court, 1993); J. Lennart Berggren and Alexander Jones, *Ptolemy's Geography: An Annotated Translation of the Theoretical Chapters* (Princeton, NJ: Princeton University Press, 2000); Andrew Barker, *Scientific Method in Ptolemy's* Harmonics (Cambridge: Cambridge University Press, 2001).

Hippocrates: Jacques Jouanna, *Hippocrates*, trans. M. B. DeBevoise (Baltimore, MD: Johns Hopkins University Press, 1999).

Galen: *Galen: Selected Works*, trans. P. N. Singer (Oxford: Oxford University Press, 1997); Vivian Nutton, *Ancient Medicine*, 2nd edn (London: Routledge, 2013); Susan P. Mattern, *The Prince of Medicine: Galen in the Roman Empire* (Oxford: Oxford University Press, 2013); Christopher Gill, Tim Whitmarsh, and John Wilkins, eds, *Galen and the World of Knowledge* (Cambridge: Cambridge University Press, 2009); Galen's medical weeks: Stephan Heilen, 'Galen's Computation of Medical Weeks: Textual Emendations, Interpretation History, Rhetorical and Mathematical Examinations', *SCIAMVS: Sources and Commentaries in Exact Sciences* 19 (2018): 201–79; Kassandra Jackson Miller, 'From Critical Days to Critical Hours: Galenic Refinements of Hippocratic Models', *TAPA* (Society for Classical Studies) 148 (2018): 111–38.

Astrology: Tamsyn Barton, *Ancient Astrology* (London: Routledge, 1994); Joanna Komorowska, 'Astrology, Ptolemy and *technai stochastikai*', *MHNH: Revista Internacional de Investigación sobre Magia y Astrología Antiguas* 9 (2009): 191–203; Dorian Gieseler Greenbaum, 'Arrows, Aiming and Divination: Astrology as a Stochastic Art', in *Divination: Perspectives for a New Millennium*, ed. Patrick Curry (Farnham: Ashgate/Gower, 2010), 179–209.

Chapter 11: Going by the book—or not

Proclus, *A Commentary on Euclid's* Elements, trans. Glenn R. Morrow (Princeton, NJ: Princeton University Press, 1970); Richard Sorabji, ed., *Aristotle Transformed: The Ancient Commentators and Their Influence*, 2nd edn (London: Bloomsbury Academic, 2016); G. W. Most, ed., *Commentaries—Kommentare* (Göttingen: Vanderhoeck and Ruprecht, 1999); Miira Tuominen, *The Ancient Commentators on Plato and Aristotle* (Stocksfield: Acumen, 2009).

Chapter 12: Beyond antiquity

Werner Heisenberg, *Physics and Philosophy: The Revolution in Modern Science* (London: George Allen & Unwin, 1958); Erwin Schrödinger, *Nature and the Greeks* (Cambridge: Cambridge University Press, 1954). In 2021, Didier Queloz, awarded the Nobel Prize in Physics in 2019, launched the Cambridge Initiative for Planetary Science and Life in the Universe, enabling cross-disciplinary research on the origin and nature of life in the universe as well as planetology (<https://www.iplu.phy.cam.ac.uk/>).

Index

For the benefit of digital users, indexed terms that span two pages (e.g., 52–53) may, on occasion, appear on only one of those pages.